Policymaking in a Nuclear Program

Policymaking in a Nuclear Program

The Case of the West German Fast Breeder Reactor

Otto Keck

LexingtonBooks
D.C. Heath and Company
Lexington, Massachusetts
Toronto

Library of Congress Cataloging in Publication Data

Keck, Otto.
  Policymaking in a nuclear program.

  Includes index.
  1. Atomic energy policy—Germany, West. 2. Liquid metal fast
breeder reactors. I. Title.
TK9203.B7K42   333.79'24'0943   79-3831
ISBN 0-669-03519-x

Copyright © 1981 by D.C. Heath and Company

Published simultaneously in Canada

Printed in the United States of America

International Standard Book Number: 0-669-03519-x

Library of Congress Catalog Card Number: 79-3831

To Monika

Contents

List of Figures

List of Tables

Foreword

A significant issue of public policy has recently emerged in the United States that will become of increasing importance in the years immediately ahead. The federal government loan guarantee to the Chrysler Corporation, more far-reaching demands for a national assistance package to help the ailing American automobile industry, and the enactment of a multibillion-dollar government-supported synfuel industry raise the pressing question of the future relationship of the government and the private sector in the United States. The ultimate answer to this question will profoundly shape the future of the American political and economic systems.

What is taking place in the United States is a local variation of a much larger development occurring in all advanced industrialized economies. The global redistribution of industry, the energy crisis, the onset of stagflation, the intensification of international competition, and the increasing importance of technological advance for economic growth and competitiveness among other developments have stimulated governments to formulate conscious and systematic industrial policies. In Japan and Western Europe these efforts are well advanced and considerable attention has been given to this subject. In the United States concern with these matters of industrial policy is just beginning and is sporadic.

The rationale for an industrial policy is that the government must intervene directly in the economy in order to change the structure of the economy. Industrial policy therefore implies a major redefinition of the relationship of the public and private sectors, at least in the case of American society. For its advocates in the United States, an industrial policy is the next logical step in the evolution of the government's role in the economy. Prior to the passage of the Full Employment Act at the end of World War II, the government's responsibility was limited to creating a legal and technical infrastructure in which private enterprise would expand. With passage of that act, the government's responsibility expanded to include stabilization of the economy through macroeconomic policies. Due to the altered circumstances of the American and world economies, the government's role, it is argued, must extend more deeply into the microlevel of the economy. It must be concerned not merely with the overall level of economic activity but also with the determination of the supply of and demand for particular goods and the overall structure of the economy.

In the developing American debate over industrial policy, considerable weight has been placed on the experiences and policies of the two most dynamic economies of our era: Japan and West Germany. Emphasis has been placed on the assistance the governments of these two economic powers seemingly provide to their industries and the unfair advantage this gives in international competition. How can we, it is argued, compete successfully against Japan Incorporated? The United States must develop, it is said, the same pattern of cooperative arrangements between business and government found in West Germany (which freely translated means business must have import protection and financial assistance). While lip service is still paid to the virtues of free enterprise and market, important sectors of American business argue that they cannot compete against such an unfair advantage.

A central aspect of this issue is the enhanced role of technological innovation in economic competition. In this situation, the increasing rate of technological innovation, the scale of modern technology, and the accelerating cost of technological developments, it is argued, necessitate government support. The risks and the costs of technological innovations are said to be beyond the capacity of the private sector. For these reasons, the government must assume the entrepreneurial role of developing new technologies, with the cooperation of American industry. An America Incorporated would be the answer to Japan Incorporated and its West German equivalent.

It is against the background of this debate over industrial policy that Dr. Keck's book is so very apposite. His thorough analysis of the German development of breeder-reactor technology is highly relevant for testing the assumption that the scale of modern technology necessitates that government assume the entrepreneurial role that in Western economic systems has traditionally been left to the private sector.

In demonstrating the fallacy of this assumption and the inefficiencies of centralized policymaking associated with large-scale government subsidies for the development of new civilian technology, he does something much more significant. His book is a necessary corrective to the perception, at least among many Americans, that the secret of German success is the intervention of the state in the economy. What he shows is that the secret of German industrial policy and economic success is reliance upon the decentralized decision-making structure of a market economy. If Dr. Keck's fine analysis of the real reason for German success helps turn us away from a growing infatuation with the concept of America Incorporated, we will be very much in his debt.

Robert G. Gilpin, Jr.
Eisenhower Professor
of International Affairs
Princeton University

Preface

Despite the intense public debate about nuclear power, many individuals on both sides agree at least on one point: that governments should support the development of nonnuclear energy technologies in a similar way and on a similar scale as they supported nuclear power. Opponents to nuclear power argue that this technology was the wrong target for government support. But they often acknowledge that government support was effective in bringing this technology to commercial application. Proponents of nuclear power have been prompted by the energy crisis to accept the view that governments should assume a greater role in developing alternative energy technologies such as solar, wind, waves, or geothermal. Among policymakers and the public it is often thought that government support to nuclear power may serve as a model for other technologies.

However, among economists opinions differ on the reasons why governments should intervene in technological innovation and on the effectiveness of different forms of government intervention. Some writers propose that public funds are required because the high costs and risks involved in the development of new technology transcend the capability even of large companies. Other authors maintain, however, that government support should be given only to the early phase of the development process, so as to establish the technical feasibility and test the specifications of new designs. If a clear economic need exists, industry will then be able and willing to develop the technology further with its own money. According to these authors, government subsidies to the later phase of the development process will be used for financing projects that firms would otherwise do with their own money or for second-best projects. It is this controversy on which this study focuses, not the public debate on nuclear power. The study seeks to contribute empirical evidence through a retrospective analysis of policymaking in the West German fast breeder program. I take no stand on the safety or environmental or proliferation hazards of the fast breeder reactor (FBR) or of other nuclear technologies. For purposes of analysis I presuppose that these hazards can be reduced to a level that is acceptable to the public.

The FBR is a type of nuclear reactor for electricity generation that has the potential to "breed" more nuclear fuel than it consumes while producing electric power. As FBRs require only a small net input of "fertile" material, usually natural uranium or waste uranium from enrichment plants, their use would considerably enlarge the world's energy resources. Governments in nearly all larger industrial countries have embarked on development programs for this technology. These programs are among the biggest ever undertaken in civilian technology. The FBR therefore provides a good case for a discussion of the rationale and effectiveness of government support for civilian research and development (R&D).

The analysis in this book covers all major decisions in the West German program: the establishment of a project group in the Karlsruhe research laboratory in 1960, the association of the program with the European Atomic Energy Community in 1963, the contracting of prototype designs to industry in 1966, the termination of industrial work on the steam-cooled fast breeder in 1969, and the construction start of a 300-MWe liquid-metal prototype in 1973. For each decision the following questions are asked: who took the initiative; what information on technical and economic matters was available to government; how did government justify the expenditure of public funds; did economic assessments made by the government prove valid in later experience; what lessons does this experience suggest as to the effectiveness of government support?

Choosing the approach of a historical case study, this analysis aims at understanding each decision in its specific historical situation within the then-existing constraints on information. Hindsight has been used to evaluate decisions, but with caution. If an economic assessment has been proved wrong by later events, this does not necessarily imply that it was bad. A good deal of judgment is required to decide if events were foreseeable or not. To avoid the subjectivity involved in such judgment, I have restricted the use of hindsight generally to instances where different actors involved in the policymaking held different views and later experience supported the views of some actors while invalidating the views of others. Events that were not expected by any of the actors involved were as a rule not drawn upon for evaluating a decision. Whether some decision strategies have better capability than others for adapting to unforeseen events is an important matter that is to be treated on its own.

Even this cautious use of hindsight forces me to say that in some instances economic assessments proved unrealistic, and in one case a decision was wrong. Such propositions may appear hard and presumptuous. It is therefore important to stress that they are meant not to question the capability or personal integrity of the individuals involved but to elucidate the organizational norms and incentives under which they acted. My criticisms do not presume that I would have made better assessments or better decisions had I acted myself. The aim of the study is to learn lessons for policymaking. Critical evaluation is a means to this end, not an end in itself.

The theoretical framework of this study is spelled out in chapter 1. Chapter 2 sets the stage through a historical account of the development of nuclear power reactors in West Germany. The main decisions in the West German fast breeder program are described in chapters 3 to 6. In chapter 7, some conclusions for future policymaking in the German fast breeder program are derived from the findings, so these may be exposed to the test of practice. Chapter 8 summarizes the findings and discusses possible generalizations.

Information was gathered from publications, from interviews with more than thirty individuals involved in the policymaking process, and from minutes and documents of government advisory bodies. Fortunately, many people in ministries, industrial firms, and government research laboratories were prepared to give information and to share their views. Some also read a draft of this book and commented on it. Most interviews were conducted from 1974 to 1977, and some in 1980. I agreed with the interviewees that they would not be identified where a piece of information is incorporated in the text. Reference is therefore made only to "interview with participants," citing the month and year. This type of reference is also used for confidential documents that I received in interviews. Most of the interviewees granted permission to include their names in the acknowledgments, so the informed reader may judge for himself the extent to which this study drew on competent sources.

According to West German regulations, documents of the federal government can be consulted for scholarly purposes before the usual thirty-year period. This requires permission of the ministry concerned. A text using confidential documents must be submitted to the ministry before publication. The ministry is entitled to comment on it and to require that these comments be published with the text. The regulations do not authorize the ministry to demand alterations in the text or to restrict publication.

This procedure was followed for this study. It is important to make clear that the permission to consult confidential documents does not imply that the Federal Ministry for Research and Technology, or any other West German government organization, or any government official, agrees with my views. Neither can such an agreement be inferred from the fact that the ministry did not wish to use its right to comment on the study and to have these comments published. Such a disclaimer applies to all individuals and organizations that kindly supplied information.

Readers wishing technical explanations can find these in any introductory text on nuclear power. For the technology of breeder reactors I found the following publications helpful: R.G. Palmer and A. Platt, *Fast Reactors* (London: Temple Press, 1961); J.R. Dietrich, "Efficient Utilization of Nuclear Fuels," *Power Reactor Technology* 6 (1963): 1–38; J.G. Yevick and A. Amorosi, eds., *Fast Reactor Technology: Plant Design* (Cambridge, Mass., and London: MIT Press, 1966); A.M. Perry and A.M. Weinberg, "Thermal Breeder Reactors," *Annual Review of Nuclear Science* 22 (1972):318–354. For recent surveys the reader may consult *Design, Construction and Operating Experience of Demonstration*

LMFBRs, proceedings of an International Symposium held in Bologna, Italy, 10–14 April 1978 (Vienna: International Atomic Energy Agency, 1978); International Atomic Energy Agency, *Status and Prospects of Thermal Breeders and Their Effect on Fuel Utilization,* Technical Report Series No. 195, (Vienna: International Atomic Energy Agency, 1979).

Acknowledgments

I am greatly indebted to the individuals interviewed, who spent their time on sometimes long or multiple sessions. Some also commented on an earlier draft of this study and discussed the findings with the author. My interpretation of facts, conclusions, and policy recommendations are not necessarily shared by the interviewees. Some, in fact, hold quite different opinions.

Interviews were conducted with D. Born (Klein, Schanzlin & Becker AG, Frankenthal); A.R. Braun (TNO, Apeldoorn); W. Breyer (Interatom, Bergisch Gladbach); R.C. Cadée (Ministry of Economic Affairs, The Hague); H. Ceulemans (CEN/SCK, Mol); J. Coehoorn (ECN, Petten); U. Däunert (West German Embassy, London; formerly BMFT, Bonn); P. Dejonghe (CEN/SCK, Mol); J. van Dievoet (Belgonucléaire, Brussels); J.A. Goedkoop (ECN, Petten); W. Gries (DWK, Hannover); E. Guthmann (Interatom, Bergisch Gladbach and Novatome, Le Plessis-Robinson); A.H. de Haas van Dorsser (Neratoom, The Hague); W. Häfele (IIASA, Laxenburg; formerly GfK, Karlsruhe); R. Hüper (KfK, Karlsruhe); G. Jünger (formerly Klein, Schanzlin & Becker AG, Frankenthal); K.-H. Kern (former member of the Bundestag, Ulm); W. Koop (SBK, Kalkar); H. Kornbichler (Uranit, Jülich; formerly AEG and KWU); D. Kuhnt (RWE and SBK, Essen); P. Lichtenberg (Secretary to the Bundestag Committee on Research and Technology, Bonn); H.-P. Lorenzen (BMFT, Bonn); H.K. Mani (Ministry of Economic Affairs, The Hague); A. Plessa (RWE, Essen and NERSA, Lyon); L. Ritz (formerly GfK, Karlsruhe); G.H. Scheuten (DWK, Hannover; formerly RWE and SBK, Essen); H. Schmale (DWK, Essen; formerly RWE, Essen); V. Schneider (Alkem, Hanau); G. Schuster (Commission of the European Communities, Brussels; formerly BMFT, Bonn); J. Seetzen (Heinrich-Hertz-Institut, Berlin; formerly GfK, Karlsruhe); K.-D. Spitz (SBK, Essen); G. Theisen (SBK, Essen); K. Traube (formerly Interatom, Bergisch Gladbach); F. Tremmel (KfK, Karlsruhe); G.V.P. Watzel (RWE, Essen); K. Wirtz (formerly GfK, Karlsruhe); W.C.L. Zegveld (TNO, Delft); A. Ziegler (University of Bochum, Bochum; formerly Siemens, Erlangen).

I am grateful to the Federal Ministry for Research and Technology for permission to consult the documents of the Deutsche Atomkommission and of the Projektkomitee Schneller Brüter, to the West German Federal Archives for access to the Zwischenarchiv at Hangelar, and to Ministerialrat P. Lichtenberg for access to the minutes of the Bundestag Committee for Research and Technology.

An earlier version of this study was accepted as doctoral dissertation by the University of Sussex in 1977.

The supervision of my doctoral research by Keith Pavitt and John Surrey at the Science Policy Research Unit of the University of Sussex was a decisive help. A particular stimulus was Keith's skepticism about the many myths surrounding government policy for science and technology and his insistence on separating opinion from empirical evidence. Thanks are due to Albert Wohlstetter, Henry R. Rowen, and the Russell Sage Foundation for the opportunity to present my results to the Workshop on Energy Options and Risks at the Center for Policy Study of the University of Chicago in November 1978, and to Victor Gilinsky and Robert N. Laney for their comments at this conference. I greatly appreciated the comments received in seminars at the Cornell Program in Science, Technology and Society and at the Institut d'Histoire et de Sociopolitique des Sciences of the Université de Montréal.

Roy MacLeod contributed detailed comments while he was subject chairman of history and social studies of science at the University of Sussex. I have profited from discussions with Lawrence Scheinman and Peter Katzenstein at Cornell; Lesley Cook at Sussex University; Jean-Claude Duperrin, visiting fellow at the Science Policy Research Unit in 1975; and A.J. Bogers at the TNO Special Studies Group at Apeldoorn, the Netherlands. Valuable information on specific items was obtained by correspondence with K. Cohen and R.B. Richards at General Electric's Advanced Reactor Systems Department in Sunnyvale, California; R. Krymn at the International Atomic Energy Agency in Vienna; and D. Streeton, Westcliff-on-Sea, England. Helpful comments on various drafts were received from William Walker, Raymond Day, and James Woudhuysen at the Science Policy Research Unit of the University of Sussex. Since English is not my native language, the help of Graham Thomas in matters of language and style was greatly appreciated.

For providing books and published documents on the subject, I thankfully acknowledge the assistance of the Fachinformationszentrum Energie, Physik, Mathematik (formerly Zentralstelle für Atomenergie-Dokumentation) at Karlsruhe, the Universitätsbibliothek Ulm, and the Zentralbibliothek des Kernforschungszentrums Karlsruhe. Herr Nenadovic of the Fachinformationszentrum Energie, Physik, Mathematik assembled two valuable bibliographies on the subject for me. In collecting paper cuttings pertinent to the subject, my work was much facilitated by Herr Stevens giving access to the treasures of his collection at the Zentralbibliothek of the Kernforschungsanlage Jülich, and by Frau Blondel guiding me through the press archives of the Presse- und Informationsamt der Bundesregierung at Bonn. I wish to thank David Fishlock, science editor of the *Financial Times,* for giving access to his files of articles.

I am grateful to Professors Helmut Baitsch and Ina Spiegel-Rösing for sharing freely the resources of the Abteilung für Wissenschaftsforschung of the Universität Ulm. Angela Erkert and Eva Rode typed several drafts from an often unreadable manuscript and assisted in managing the voluminous correspondence and the extensive bibliographical work. Further assistance in typing was received from Marie-Luise Hefuna, Helga Lohrmann, Rita Pieper, and Annemie Schmitz. Ursula Klaus-Hillner and Angelika Walser helped in the bibliographical work, and Max Häge in some computer calculations.

The study was supported by grants from the Deutsche Forschungsgemeinschaft and the University of Ulm. The procurement of pertinent literature was very much facilitated by a grant of the Volkswagen Foundation to the library of the University of Ulm.

Without such help by numerous people this book would not have been feasible. Of course, the mention of an individual or an organization does not imply that they approve my presentation of facts, conclusions, or policy recommendations. The responsibility for these, including shortcomings and errors, remains on my shoulders.

List of Abbreviations

AEC	U.S. Atomic Energy Commission
AEG	Allgemeine Elektricitäts-Gesellschaft
AGF	Arbeitsgemeinschaft der Grossforschungseinrichtungen
AGR	Advanced gas-cooled reactor
AKS	Arbeitsgemeinschaft Baden-Württemberg zum Studium der Errichtung eines Kernkraftwerks (in short, Arbeitsgemeinschaft Kernkraftwerk Stuttgart)
ANS	American Nuclear Society
AVR	Arbeitsgemeinschaft Deutscher Energieversorgungs-unternehmen zur Vorbereitung eines Leistungs-Versuchs-Reaktors e.V. (since 1959, Arbeitsgemeinschaft Versuchs-Reaktor-GmbH)
BBC	Brown, Boveri & Cie AG (Mannheim)
BBK	Brown, Boveri/Krupp Reaktorbau GmbH
BBR	Babcock-Brown Boveri Reaktor GmbH
Benelux	Belgium, the Netherlands, and Luxemburg
Bewag	Berliner Kraft und Licht AG
BMFT	Bundesministerium für Forschung und Technologie, Bonn
BWR	Boiling-water reactor
CDFR	Commercial Demonstration Fast Reactor (Great Britain)
CEA	Commissariat à l'Energie Atomique (France or Belgium)
CEGB	Central Electricity Generating Board (Great Britain)
CEN/SCK	Centre d'Etude de l'Energie Nucléaire/Studiecentrum voor Kernenergie (Mol, Belgium)
DFR	Dounreay Fast Reactor (Great Britain)
DM	Deutschmark
DPf	Pfennig (a hundredth of a deutschmark)
DWK	Deutsche Gesellschaft für Wiederaufbereitung von Kernbrennstoffen mbH
ECN	Stichting Energieonderzoek Centrum Nederland (Netherlands Energy Research Foundation)
EDF	Electricité de France
ENEA	European Nuclear Energy Agency of the Organisation for Economic Co-operation and Development (since 1972, Nuclear Energy Agency)
ENEL	Ente Nazionale per l'Energia Elettrica (Italy)
ESADA	Empire State Atomic Development Associates, Inc. (United States)

ESK	Europäische Schnellbrüter-Kernkraftwerksgesellschaft mbH
Euratom	European Atomic Energy Community
FBR	Fast Breeder Reactor
FDP	Freie Demokratische Partei
FFTF	Fast Flux Test Facility
FR 2	Research reactor at Karlsruhe
g	Gram
GCFBR	Gas-cooled fast breeder reactor
GE	General Electric Company (United States)
GfK	Gesellschaft für Kernforschung mbH (Karlsruhe; since 1978, Kernforschungszentrum Karlsruhe)
GHH	Gutehoffnungshütte Sterkrade AG
GKSS	Gesellschaft für Kernenergieverwertung in Schiffbau und Schiffahrt mbH
GWe	Gigawatt (electric)
HDR	Heissdampfreaktor
HRB	Hochtemperatur-Reaktorbau GmbH
HTGCR	High-temperature gas-cooled reactor
IAEA	International Atomic Energy Agency
IIASA	International Institute for Applied Systems Analysis
INB	Internationale Natrium-Brutreaktor-Bau-Gesellschaft mbH
INFCE	International Nuclear Fuel Cycle Evaluation
KfK	Kernforschungszentrum Karlsruhe
kg	Kilogram
KKN	Kernkraftwerk Niederaichbach
KNK	Kompakte Natriumgekühlte Kernreaktoranlage
KRB	Kernkraftwerk RWE-Bayernwerk GmbH
kW	Kilowatt
kWe	Kilowatt (electric)
kWh	Kilowatt-hour
KWL	Kernkraftwerk Lingen GmbH
KWO	Kernkraftwerk Obrigheim GmbH
KWU	Kraftwerk Union AG
LMFBR	Liquid-metal fast breeder reactor
LWR	Light-water reactor
MAN	Maschinenfabrik Augsburg-Nürnberg AG
MW	Megawatt
MWe	Megawatt (electric)
MWsec	Megawatt-second
MWt	Megawatt (thermal)
MZFR	Mehrzweckforschungsreaktor

NACA	National Advisory Committee on Aeronautics
NEA	Nuclear Energy Agency of the Organisation for Economic Co-operation and Development
NERSA	Centrale Nucléaire Européene à Neutrons Rapides S.A.
OECD	Organisation for Economic Co-operation and Development
ORNL	Oak Ridge National Laboratory
PFR	Prototype Fast Breeder (Great Britain)
PSB	Projektgesellschaft Schneller Brüter GbR
Pu	Plutonium
PWR	Pressurized-water reactor
R&D	Research and development
RCN	Stichting Reactor Centrum Nederland
RWE	Rheinisch-Westfälisches Elektrizitätswerk AG
SAEA	Southwest Atomic Energy Associates (United States)
SBK	Schnell-Brüter-Kernkraftwerksgesellschaft mbH
SCFBR	Steam-cooled fast breeder reactor
SEEN	Syndicat d'Etude de l'Energie Nucléaire (Belgium)
SEFOR	Southwest Experimental Fast Oxide Reactor
SEP	N.V. Samenwerkende Electriciteits-Productiebedrijven (Netherlands)
SKW	Studiengesellschaft für Kernkraftwerke
SNEAK	Schnelle Null-Energie-Anordnung Karlsruhe
SNR	Schneller Natriumgekühlter Reaktor
SST	Supersonic transport
STARK	Schnell-Thermischer Argonaut-Reaktor Karlsruhe
SUAK	Schnelle Unterkritische Anordnung Karlsruhe
t	Tonne (metric)
TNO	Organisatie voor Toegepast Naturwetenschappelijk Onderzoek (Netherlands)
TNPG	The Nuclear Power Group Ltd.
U	Uranium
UKAEA	United Kingdom Atomic Energy Authority
VAK	Versuchsatomkraftwerk Kahl
VEW	Vereinigte Elektrizitätswerke Westfalen AG
WAK	Wiederaufbereitungs-Anlage Karlsruhe

1

Government Support for Civilian Technology

Inspired by the success of large-scale projects in military technology during World War II, in particular the Manhattan project producing the atomic bomb, governments in the United States and in other advanced capitalist countries have increased considerably their support for the development of new technologies. In addition to a rapid expansion of government funds for research and development (R&D) in the military sector, support has been given on a growing scale for civilian technology. Large-scale government support for civilian technology has now become an accepted practice in many advanced capitalist countries, and it is being directed at an increasing number of industrial sectors.

Growth of Government-Financed R&D

While in 1940 the U.S. government spent $74 million on R&D,[1] this expenditure rose to $2.8 billion in 1953 and to $21.8 billion in 1977 (see table 1-1). In West Germany, public expenditure on R&D grew in the period 1953–1977 from DM0.8 billion to DM12.8 billion. Support for civilian R&D (excluding space) increased even faster than for military R&D: in the United States from $0.2 billion in 1953 to $7.5 billion in 1977, in West Germany from DM0.8 billion in 1953 to DM10.6 billion in 1977. If the influence of inflation is excluded by expressing expenditure in constant prices (see table 1-1), it can be seen that growth has been considerable in real terms. While in the United States total government R&D decreased from the mid-1960s to the second half of the 1970s as a result of cutbacks in military and space technology, expenditure on civilian R&D grew continuously in real terms in the United States as well as in West Germany.

The motives for expansion in government expenditure for civilian R&D were manifold. In nuclear power, governments have taken great efforts in transferring to civilian uses the scientific and technical knowledge acquired in wartime and postwar military programs. In some technologies, such as aircraft and electronics, the (real or imagined) side benefits of military projects to the civilian sector of the American economy induced governments in other countries to embark on large civilian programs in order to catch up with a (real or imagined) American lead in technology. Events on the world market for energy raw materials since 1973 have contributed considerably

1

Table 1-1
Expenditure on Research and Development in the United States and West Germany

Source of Funds Government Objectives [a]	United States			West Germany		
	1953	1965	1977	1953[b]	1965	1977
	Current prices (billion dollars or deutschmark)					
Government	2.8	13.0	21.8	0.8	3.7	12.8
Defense and space	2.6	10.8	14.3	0.0	0.8	2.2
Civilian	0.2	2.2	7.5	0.8	2.9	10.6
Industry	2.2	6.5	19.6	0.4	4.1	14.3
Other (incl. abroad)	0.1	0.5	1.5	0.0	0.1	0.9
Total	5.1	20.0	42.9	1.2	7.9	28.0
	1972 prices [c] (billion dollars or deutschmark)					
Government	4.7	17.5	15.4	1.4	5.1	9.9
Defense and space	4.4	14.5	10.1	0.0	1.2	1.7
Civilian	0.3	3.0	5.3	1.4	3.9	8.2
Industry	3.8	8.8	13.8	0.8	5.5	11.0
Other (incl. abroad)	0.2	0.7	1.1	0.0	0.1	0.7
Total	8.7	27.0	30.3	2.2	10.7	21.6
	Percentage of total national R&D					
Government	54	65	50	64	47	46
Defense and space	49	54	33	0	11	8
Civilian	5	11	17	64	36	38
Industry	44	33	46	36	52	51
Other (incl. abroad)	2	2	4	0	1	3
Total	100	100	100	100	100	100

Sources: National Science Foundation, *National Patterns of R&D Resources 1953-1980* (Washington: Government Printing Office, 1980); H. Echterhoff-Severitt et al., *Forschung und Entwicklung in der Wirtschaft 1975* (Essen: Stifterverband für die Deutsche Wissenschaft, 1978); OECD, *Changing Priorities for Government R&D* (Paris: OECD, 1975); *Bundesbericht Forschung VI, Bundestagsdrucksache* 8/3024 (Bonn: Heger, 1979).

[a] Inconsistencies are due to rounding or to inconsistencies among different sources. Money amounts in "defense and space" and "civilian" were derived from percentages and may therefore include small margins of error.

[b] Figures for West German expenditure in 1953 include academic education in addition to R&D.

[c] Based on GNP price index.

to the responsibility of governments for civilian technology, as technological innovation is expected to help energy-importing countries to limit the political and economic consequences of dependence on foreign oil supplies. This list is not exhaustive. In general, civilian R&D has been increasingly recognized to make a significant contribution to various economic, social, and political goals and thus has received an increasing proportion of governments' attention and their budgets.

Compared with government support for civilian technology before World War II, postwar policies mark a change not only in scale but also in relations between government and business enterprises. In Western capitalist countries there has been a long tradition of government support in areas such as agriculture, geography, fundamental natural science, and military technology. This comprised support for government research laboratories and, in European countries, for university research, but a good deal of government support was ad hoc and piecemeal. While part of the growing government R&D after World War II was geared to traditional areas of government activity, an increasing part was intended to influence the activity of civilian industry in a direct manner: through contracts, through subsidies, or through the production of knowledge in government laboratories for direct use by industrial firms.

In some countries and some sectors, government R&D was associated with the nationalization of a technology. Britain and France followed such a policy for some time in civilian nuclear power. However, in most sectors of advanced technology the increased role of government did not suspend the traditional division of labor between a public sector and a private sector, though it did effect significant changes within this framework. The increase in government R&D has not diminished the importance of the business enterprise sector as a source of R&D expenditure. The self-financed R&D expenditure of the business enterprise sector has grown at similar rates as government expenditure (see table 1-1). American industry increased its self-financed R&D expenditure from $2.2 billion in 1953 to $19.6 billion in 1977, West German industry from DM0.4 billion in 1953 to DM14.3 billion in 1977. In 1977, American industry contributed 46 percent of national R&D funds; in West Germany, industry's contribution was 51 percent.

R&D is thus part of the division of work between government and the business enterprise sector that is typical for the modern mixed economy. Production of goods and services is divided between two different systems of public control: the political process that in theory is governed by the citizen's vote, and the market that in theory is governed by the consumer's choice. In reality, both control mechanisms are more complex, as many forces mediate between the citizen's vote and government decisions, and between the consumer's choice and producer decisions.

As for any economic function in a society, a basic question concerning R&D is which function is to be allocated to which system of control. Despite a relative consensus that in the real world both systems of control work more or less well if compared with their ideal, this question often rouses political controversy. And if it is treated on a rather general level, this controversy is usually loaded with ideological dimensions. Different systems of thought have developed different rationales for an appropriate division of labor between government and the business sector and the appropriate mix

of political and market controls. There are two basic approaches for establishing such a rationale. One starts with a centralized political system and
derives from its weaknesses the role to be left to business enterprises and the
decentralized control of market forces. The other begins with a decentralized market and derives from its weaknesses the role to be left to government and central political control. The following discussion of rationales of
government support to R&D moves along the latter approach, as it is closer
than the first to the historical way in which government support to R&D has
developed in advanced capitalist countries. This approach commends itself
for a case study that wants to understand its subject in its precise historical
context, rather than measure it against abstract normative theory.

Rationales of Government Intervention

The economic framework of advanced capitalist countries is guided by the
principle that goods and services are provided primarily by business enterprises through the market. Competition and profit are to create pressures
and incentives for firms to produce goods and services, and the market is to
match supply with demand. Governments have in a capitalist framework
the task of providing services and goods which the market cannot supply
adequately. Typical public-sector functions are defense, general education,
public health and safety, security, certain kinds of transportation and communication, social welfare, weights and measures, and weather forecasting.
Furthermore, governments have taken on the responsibility for fostering
the prosperity of the economy in order to maintain full employment and
economic growth. This responsibility is disposed of by overall regulation of
the business cycle, international economic policy, tax incentives, and other
means. While some of these measures aim at the economy as a whole, others
are directed to specific sectors of the economy.

In general, the same principles also guide technological innovation and
R&D.[2] The forces of the market drive firms to develop and deploy new
products and processes and to look for improvements in existing products
and processes. R&D is an essential element in the innovative activities of
firms in most industrial sectors. However, when specific services are not
provided by market forces and government has to take care of them, market
forces are also unlikely to bring forth the scientific and technological
knowledge which may help to perform these services better. Although the
government may purchase many inputs for these services from business
firms which themselves do R&D on their products and services, this will not
produce knowledge for those aspects of government activity that are
beyond the direct interests of supplier firms. A good deal of R&D for public-sector functions will therefore have to be financed by government.

Science and technology also have a significant role in government efforts to maintain or increase economic prosperity. In modern industrial society, technical progress has become a precondition of economic growth and international economic competitiveness. A country's material wealth and political influence on the international scene depend upon how effectively its economy is putting technology to use.[3] Although the allegiance of the population is contingent not only upon material wealth produced by technological advance but also, and to an increasing extent, upon how effectively governments control undesired consequences of new technology, the technological capability of domestic industry does remain a matter of national concern.

The need for government intervention in civilian technology outside the public sector is generally justified by imperfections of the market mechanism.[4] Business enterprises directed by the pressures of competition and by the incentives of profit will not bring forth a supply of new civilian technology that is adequate in relation to the benefits which the economy as a whole and the general public are to derive from it. The investments of a firm in innovative activities will reflect the returns it expects from a new technology. The market price that is the main determinant of these returns will not adequately reflect all the costs of potential undesired consequences of the new technology, or all the possible benefits that the economy or the general public may derive from the new technology. As governments in many cases take measures to limit external costs of new technology, for instance by imposing pollution standards, so they promote industrial technology in order to create social benefits that would be foregone if the industrial system was left on its own.

While the divergence of private costs and benefits from social costs and benefits may exist in various kinds of economic activity, a special argument may be made for industrial R&D. Scientific and technical knowledge, unlike other goods, is not diminished in its value when it is used. Although a firm may exclude others from using some of its know-how by such measures as secrecy, it may not be able to control all its know-how when it brings a new product on the market. Other firms may imitate the product, though this may require some R&D on their own. The possible use of the scientific and technological knowledge by other firms may reduce the benefits accruing to the performer of R&D, and thus may reduce its incentives to invest in R&D.

The divergence of social from private costs and benefits may provide a persuasive rationale in theory, but in reality it yields limited guidance for identifying market imperfections and potential corrective action.[5] Compared with the ideal world of neoclassical economics, there are many market imperfections that affect decisions of business enterprises, and the combined impact of these imperfections on the propensity of firms to finance R&D and to innovate is difficult to discern. Some imperfections may work toward a higher rate of technological change than optimal for society. Oligopolistic markets with competition directed at product improvements rather than prices may be such a case. Externalities cannot be routinely measured, so the divergence of social from private costs and benefits does not provide a neat calculus for an optimal government intervention in civilian technology.

Largely by common sense and unsystematic trial and error, government practice has developed an arsenal of measures for fostering industrial innovation. On an intermediate level, economists have provided justifications for these measures. These yield broad indications for what governments should do, but no precise criteria for government action. Thus, there is scope for political controversy and scholarly discussion.

All governments have a patent system. This has been justified by the increased incentives that the excess benefits of a temporary monopoly provide to the innovating firm. Some governments have special tax rates for industrial R&D. This increases the benefits to the innovator by reducing his cost. All governments provide funds for basic research. Economists have provided a rationale for this by pointing to the difficulties of predicting the practical benefit that may result from a piece of basic research, and to the difficulties that performers of basic research have in appropriating these benefits.[6] Governments have traditionally financed R&D in the agricultural sector. This may be justified by the fact that productive units in this sector are too small for performing R&D. While manufacturing firms supplying goods and services to farmers do R&D that is useful to farmers, this does not adequately cater for those aspects that are outside the immediate interests of suppliers.

What is the rationale for large-scale government support to R&D in modern sectors of industrial technology? What are the factors preventing business enterprises in these sectors from providing an adequate output of new technology? Governments and economists have derived a rationale from a number of arguments. But these have been challenged by some writers, especially liberal economists.[7] Three main arguments are presented, together with their counterarguments.[8]

First, it is often said that development of new technology involves such large costs, such high risks, and such long lead times, that it is beyond the financial capability of business enterprises. The counterargument maintains that business enterprises are capable and willing to make large investments in the development of new products or processes, provided there are good prospects for a satisfactory economic return. Government money will be used by firms to replace their own funds, or it will be spent on second-best projects with poor economic prospects. Instead of putting government money into the costly final stages of the development process that are aimed at bringing a specific design to commercial use, this money would be better used to finance fundamental research and exploratory development. For large prototype and demonstration plants government loans and risk sharing should be sufficient.

Second, the role of technology in international economic competition has been used to justify large-scale government support. As a country's competitive position on international markets depends on the technological capability of domestic industry, a technological lead of domestic industry on world markets may be important for sustaining a favorable balance of trade. Equally a technological lag may severely encroach upon a country's economic performance. Government support then is necessary to help domestic industry retain its lead or catch up with foreign industry. The writers challenging this argument point out that any country will command only a minor part of the world's technological capability, so there will always be some technological sectors where other countries are ahead. The conventional economic wisdom that a country will do better by exploiting comparative advantages than by striving for autarky applies also to technology, so it is far from being optimal to build up a technological capability across the whole front. Within the framework of international exchange, the premium is for specialization. If there is a technological gap with regard to other countries, which is to be diminished (if other countries are ahead) or to be sustained (if other countries are trying to catch up), governments should back industry by financing the infrastructure and a capability for exploratory development in the specific technologies rather than by committing themselves to subsidies for full-scale commercial development.

Third, a variant of the previous justification focuses on the problems of security of supply that are inherent in international trade. Certain sectors of technology have a key role in the modern industrial system, as they have an important impact on productivity and competitiveness in other sectors. Reliance of a country on foreign sources of supply for these technologies may become dangerous, as foreign governments or foreign firms holding a monopoly may threaten to cut off supplies for political or economic objectives. This reasoning has been called into question by writers who argue that foreign supply of key technologies does not call for building up a domestic capability broad enough to challenge the foreign supplier over the whole front of the commercial market. Industry, backed by a government-financed exploratory R&D, should be allowed to follow the lines it deems most profitable, and government funds should be used to maintain exploratory R&D in the critical technology as a sort of insurance policy, so domestic production is possible in case a monopoly is used for economic or political power.

What evidence can empirical research contribute in order to resolve these differences of opinion? There is some empirical knowledge on the costs and the lead times in industrial innovation, probabilities of success and failure, and also on the impact of new technology on international trade. And this evidence may lead us some way in discussing the rationale of government support to industrial innovation. However, as in other areas of government intervention, conclusive evidence could be produced only by comparing the activities of firms under different regimes of government support. This runs counter to serious conceptual and practical problems. Such a comparison may be achieved by an experimental approach. This would require a larger number of government-financed projects and would be possible only in such sectors of technology where costs of each project are small. By definition, such an approach is not practicable for technologies where costs are high, and thus would make no contribution to a discussion of the first argument. Difficulties of a conceptual kind arise from the fact that government support is designed to influence the behavior of firms. Once a certain kind of government support is given, we can only speculate on how firms would have behaved under a different regime of government intervention or with no support at all. Another problem is associated with the notions of success and failure in technological innovation. The success of a government-financed innovation does not necessarily prove the success of government intervention, as a firm might have done the same project with its own funds in the absence of government aid. Vice versa, the failure of a government-financed project does not necessarily invalidate government policy, as there are failures also in self-financed R&D.

This study attempts to circumvent these difficulties by looking at the process of policymaking. Differing views on the need for and the effectiveness of large-scale government support to industrial technology are guided by differing sets of assumptions: on the nature of technological innovation and its economic impacts; on the behavior of firms without government support; on the nature and sources of inadequacies in the firms' output in relation to external benefits which the government deems desirable for certain economic and political objectives; on the processes by which governments identify these inadequacies; and on the effects of corrective action on the behavior of firms. In the policymaking process there is an interaction between government and industry, and analysis of the policymaking process may enable us to discern to which set of assumptions the observed behavior conforms best. Of course, a historical case study analyzing only a few policy decisions can provide only limited evidence. The behavior of the actors may be influenced by idiosyncracies of individuals or by historical accident. The concluding chapter of this book discusses the extent to which these specific factors can be sorted out and offers some tentative generalizations derived from the findings.

To elaborate the framework of analysis, two different views of government support to industrial R&D with two different sets of behavioral and economic assumptions are now described in more detail. One is taken from the writings of John Kenneth Galbraith, the other from the writings of Richard R. Nelson and George Eads, with some occasional references to additional writers.

Imperative of Technology?

The starting point for Galbraith is the high risks inherent in modern technology.[9] To develop a new technology and put it to commercial use requires heavy investments, which have to be made in great uncertainty with regard to technical and economic factors. Technical failures or a misjudgment about market demands can engender heavy losses. Committing large funds under high uncertainty is exactly what defines high risk. Galbraith illustrates this by an example from automobile production. When the Ford Motor Company was founded in 1903, negotiations to form the company took only a few months. After the factory was established, only a few more months passed till the first car reached the market. In the 1960s, however, preparations for a new model required three and a half years.

Galbraith attributes this difference to the nature of modern technology. In the manufacture of the first Ford automobiles, metallurgy for instance played no part within the factory. Ordinary steel was used, which could be bought in the morning from the warehouse and manufactured the same day. The provision of steel for a modern vehicle, however, includes the elaboration of specifications by the engineer in the laboratory, preparation of appropriate metalworking machines, testing procedures, and so on. This is the reason why the application of technology leads to an *increased span of time* between the decision to produce and the first output of a salable product.

As a second consequence of modern technology, Galbraith singles out the *increase in the capital* that is committed to the production process. When Ford Company was founded, the issued stock amounted to $100,000. Introducing a new model in 1964 cost $9 million for engineering and styling, and $50 million for investment in new machinery and adaptation of existing machinery.

A third consequence is the *specialization* of machinery, materials, and manpower. The machines used in the fabrication of the first Ford motorcars were not specialized on specific components. They could be adapted to the production of other goods like bicycles or steam engines in a few hours. In the case of a modern car, however, the production machinery is designed in a specific way and cannot be used for the production of a different model without changes involving considerable time and cost. Technology requires specialization also with regard to manpower, as new products draw upon the specialized knowledge of many people with expertise in very specialized matters.

The longer the time elapsing between the decision to produce and the sale of the product on the market, the more uncertain are anticipations of market demand which guide the decision to produce. The more specialized the product, the more inelastic is market demand, and the greater are the sanctions for erroneous anticipations of what quantities will be salable at a profitable price. The more specialized the inputs in machinery and materials, the larger is the uncertainty that the market will supply the inputs at the time they are needed at costs consistent with a profitable price of the product. Hence, technology enhances greatly the *uncertainties of the market*. As heavy investments are at stake, these uncertainties engender a high risk.

Galbraith thinks that the risks are the greater, the more sophisticated the technology. Sophisticated technology implies longer lead times, higher cost, more specialization, and a lot of technical problem solving. In more intricate technologies, such as aircraft, space vehicles, nuclear power, and high-speed ground transport, new products require large efforts in *research and development*. This engenders high costs in time and money and adds a peculiar uncertainty as to whether technical problems will be solved.

There are various ways open to the corporation to reduce risks. The specialization of manpower and machinery has its counterpart in *large and skillfully designed organization,* whose function is to coordinate the work of specialists and to bring it to a coherent result in order to reduce the uncertainty of the market. In addition to skillful market research and testing, the firm will try to minimize or to get rid of market influences by long-term contracts between producers and purchasers or by taking over supplier firms or customer firms. As the capability of a firm to influence the market is a function of its size, Galbraith appreciates *big size* of corporations not as a danger to competition, but as a prerequisite for the corporations to cope with the market uncertainties engendered by technology. In addition, a large firm can bear commercial failures which a smaller firm would not be able to survive. Size, for Galbraith, is thus not a result of profit seeking, but "the general servant of technology."[10]

However, Galbraith identifies some features in the organizational structure of the modern corporation which make its management peculiarly averse to technical risks. Modern technology makes decision making so complex, that it cannot be done by a few individuals at the top. As the expertise of many different specialists must be brought to bear on decisions, decision making is done by committees rather than by individuals, and it involves many people below the rank of the top executives. For the people participating in the decision making of the corporation, Galbraith has coined the term *technostructure*. The technostructure has different goals and different attitudes toward risk than the former entrepreneurial firm. For the entrepreneur, maximization of profits is the primordial goal of his business activity. Profits or losses accrue to the person who makes the decisions. This is different in the modern corporation, where profits accrue not to those in charge of decision making but to stockholders.

For the technostructure, as for any organization, the primary goals are survival and autonomy. These goals require a level of earnings large enough to pay accustomed dividends to stockholders and to provide sufficient funds for reinvestment. As long as these requirements are fulfilled, the technostructure is secure and safe from adverse interference from stockholders or banks. Inadequate return invites investigation and intervention from stockholders and banks. This may encroach upon the autonomy of decision making of the technostructure. Prevention of loss therefore carries highest importance for the technostructure. Whereas the entrepreneur may risk low profits for a short time if he sees a good chance to reap thereafter extraordinary revenues, the technostructure places a premium on avoidance of risk.

Once survival and autonomy are ensured by a minimum level of earnings, the greatest possible rate of corporate growth is the premium goal of the technostructure. Corporate growth means expansion of the technostructure, more jobs, promotion, and higher pay. Maximum corporate growth rather than maximum profits commends itself to the self-interest of the technostructure, as maximum profits accrue to stockholders, not to the technostructure.

After a minimum level of earnings and of corporate growth are achieved, the technostructure has scope to follow other organizational interests and may seek technical progressiveness for its own sake, or, using Galbraith's term, "technological virtuosity."[11] A high competence in technology is functional to corporate growth, as improvements of products help a firm to hold existing markets and new products help to capture new markets. In addition, technological progressiveness is serviceable to the interests of the technostructure, as it implies more jobs and promotion for technologists. Achievements in sophisticated technology add to the prestige of individual members of the technostructure and enhance the image of the corporation.

Technological progressiveness can be a goal only as far as it does not interfere with the superior goals of the technostructure. It is particularly the goal of a secure minimum profit that places limits on the corporation's technological virtuosity. This pressure to avoid risks combined with the technostructure's interest in technological progressiveness are identified by Galbraith as the forces which make the modern corporation seek government support for R&D. Although these forces are expressions of organizational self-interest, this self-interest is a servant to modern technology. In

Galbraith's own words: "Technological compulsions, and not ideology or political wile, will require the firm to seek the help and protection of the state."[12] If society does not want to forego the benefits of modern technology, the state must bear the cost. Government support is therefore an imperative of technology.

In dealing with specific areas of government support to industrial technology, military technology is Galbraith's focus, reflecting the priorities of government in the United States at the time he wrote his book. The defense budget of the U.S. federal government accounted then for roughly half of the federal expenditure. Galbraith attributes to military expenditure the function of underwriting technology. Contracts for military supplies provide a nearly riskless branch of activity, so firms can take risks in their civilian activities which would otherwise be unacceptable. As the bulk of the military budget is spent on sophisticated technology, it gives firms access to technological areas from which they would otherwise be excluded by cost and risk. It pays for technological advance that may be useful also in the civilian sector.

Since the weapons competition between the United States and the Soviet Union is not without danger of an eventual war that could incinerate whole continents, Galbraith urges for a policy of disarmament and considers functional alternatives to the weapons economy of the Cold War. One such alternative is the space race. Provided that scale and technical sophistication are equivalent to weapons, other fields of technology are also suitable targets for government support such as transport airplanes, supersonic aircraft, high-speed ground transport, nuclear energy, exploration of the ocean floor, drilling through the earth's crust, and weather modification.

Galbraith provides some interesting assumptions on the interaction of industry and government in the policymaking process. In sectors of advanced technology, members of state and business organizations work together in the decision-making process. Firms take an active role in defining government objectives and in determining technological specifications in government procurement. The links between a government agency and its contractor firms may be similar to those between the agency and its superior authority. Hence the line dividing state and private business has become increasingly blurred.

Although the dissolution of the boundary between state and private business opens the possibility for influence in both directions, Galbraith sees the predominant influence on the part of industry. What is promoted by the state as national goals reflects in fact an adaptation of public goals to the needs of the industrial system. The state's concern for a stable economy, economic growth, national defense, and national technological excellence has its counterpart in the needs and goals of the technostructure. The technostructure requires economic stability as well as military and space procurement for its planning; economic growth is conducive to its organizational needs; and it relies on the state for the advance in technology. In addition to participation in government decision making, the construction of appropriate public images is described by Galbraith as an effective mechanism for industry to influence public goals. By promoting the image of a nation being threatened by an enemy, industry may produce a response of the political system in favor of increased military procurements. By picturing a technological lag relative to other countries, it may ensure increased government spending on R&D.

Subsidy to the Second Best?

Eads and Nelson put forward their view in a discussion of the U.S. programs for the fast breeder reactor (FBR) and civilian supersonic transport (SST).[13] Whereas proponents of these programs have referred to the argument that the scale of development costs, the length of lead times, and the high risks involved transcend the capability of private firms, these authors propose that the real reasons why American industry has not engaged in these projects were poor commercial prospects and an expected rate of return that was much lower than from other uses of resources. They maintain that firms do invest huge sums in R&D projects, when they can expect a satisfactory return. As evidence for this, they refer to IBM's investment in the development of the System-360 family of computers, which is reported to have amounted to $5,000 million.[14] In relation to sales, total assets, and net worth, this is more than the estimated costs of SST development in relation to the potential of the contractors for the SST. Furthermore, they point out that firms do engage in long-term projects with uncertain and distant returns, as shown by industrial R&D on television and polymers in the 1920s.

Government sponsorship may be called for because private profit opportunities do not reflect social benefits. But Eads and Nelson emphasize strongly that traditional U.S. government policy steered shy of subsidies to specific commercial projects through the later phases of development, which were left to private business. As an example of traditional government support of civilian technology, they point to the policy of the National Advisory Committee on Aeronautics (NACA) in the 1930s. NACA focused on development and operation of facilities for general use such as wind tunnels, information services, and financial support to basic research and exploratory development. The function of exploratory development is defined by Eads and Nelson as testing the broad specifications of new designs and demonstrating technical feasibility. A clear distinction is made between this kind of development activity and a development effort aimed at launching a specific product on the market. The latter activity they call commercial development. The decision to bring specific designs to commercial application has traditionally been left to business enterprises that financed the activity from their own funds.

Even with government backing of exploratory R&D, some projects requiring large financial resources and long lead times may be beyond the capabilities of private firms and may call for some public action. But in this case, contrary to projects with considerable uncertainties about economic returns, loans and risk sharing by the government should be sufficient.

Eads and Nelson stress that there was no economic need for nuclear power and civilian SST at the time initiatives were taken. Energy supply was ample in the 1950s when civilian nuclear power was promoted. Adequate supply seemed assured at least up to the turn of the century. As for civilian SST, there are signs that development of new civilian aircraft in the postwar period was faster than was economically desirable. Proponents of civilian SST and nuclear power framed their arguments not in terms of economic problems or needs to be responded to, but in terms of technological opportunities to be seized. That those technological opportunities should be exploited was not justified in terms of an investment with good returns but with reference to external social effects, such as the social benefits of sustaining American technological leadership or of having the technology available earlier than traditional government support would permit.

Proponents of civilian SST based their case on a close connection between technological leadership and commercial dominance in world markets. If technological leadership is lost, this may irreversibly impair the commercial position in world markets, which finally will result in balance-of-payment problems with undesirable effects on domestic employment and standard of living. Eads and Nelson, however, contest the assertion that technological leadership is equated with commercial dominance of the world market. Commercial success depends on technological superiority only insofar as the superiority is connected with advantages in the cost characteristics of the product. For an example, they refer to studies showing that the American aircraft industry has dominated the world market in spite of the fact that the British were first to introduce civilian jet transport.[15] What matters is not technological leadership but a commercial sense in deciding which technological advances to embody into commercial products at what time.

The misplaced emphasis on technological leadership is seen by Eads and Nelson as the result of a confusion between the characteristics of civilian and military technology. In military technology, sometimes even a small technological advantage may make the difference between life and death. But in a commercial product, technological sophistication per se has no value. The uses of a product or a process in relation to costs are the important factors.

Nelson finds that among advanced industrial nations fears about international technological leads or lags are largely without foundation.[16] He argues that, though a technological lead of the United States is a long-standing phenomenon dating back to the nineteenth century, no country is in the position to sustain a general and comprehensive leadership in every technological field. This would be an unrealistic objective as would be an effort of a country to catch up with others across the whole front of technology.

An argument for externalities in form of having the technology sooner than the market would provide, was used particularly in cost-benefit analyses for the American fast breeder program. Eads and Nelson, however, stress that speeding up the development of a technology does not only make benefits flow earlier but also increases development costs and risks. Referring to a study of government R&D in defense, they argue that there is a time-cost trade-off in technical development.[17] Furthermore, the net benefits of having a specific technology quicker than if left to private business must not be taken at its absolute value but must be compared to the net benefit from programs which will not be undertaken because of haste in the specific technology.

With regard to government-industry interaction, Eads and Nelson arrive at a view which is directly opposed to that of Galbraith. Though they argree that government subsidy to commercial development, with government and industry jointly making decisions, blurs the borderline between public and private sectors, they point out that the initiatives for civilian SST and for nuclear power came from within government and, contrary to usual views of government practice, did not result from pressure by external interest groups. The early thrust for large-scale government activities in nuclear power came from the U.S. Atomic Energy Commission (AEC) and the Congressional Joint Committee on Atomic Energy, while manufacturing industry and utilities were not enthusiastic about it. Plans for civilian SST emerged from a coalition of two government agencies: the National Aeronautics and Space Administration (NASA) and the Federal Aviation Administration (FAA).

Eads and Nelson strongly criticize the centralized institutional structure which is evolving through government support of commercial development. Contrary to Galbraith, they find this structure highly inefficient, costly, and even pernicious. The traditional decentralized framework of industrial R&D is characterized by quick and flexible decision making and a diversity of sources of technical knowledge. As early assessments of technologies are uncertain, many projects which at first looked promising later failed, while many breakthroughs were unexpected and initially did not gain the experts' support. Good ideas have a variety of paths to make themselves heard.

Firms draw information inputs for their R&D projects from many different sources: suppliers, purchasers, outside individual inventors. This process of information transfer is not orderly, but Eads and Nelson think that an attempt to impose order is likely to lead to worse results.

In the centralized structure adopted in military R&D as well as in the civilian power program of the AEC, Eads and Nelson find a tendency to stick to projects in the face of mounting evidence that commercial prospects are poor. They assume that this will happen generally when government indulges in commercial projects. Jewkes spells out in more detail the reasons for this proclivity to stick to bad projects.[18] When a government is committed to a specific project, withdrawing from the project implies confession of failure and a possible loss of face. Abandoning projects is disagreeable for all institutions concerned, and even in the absence of formal obligations, institutions depending on government money expect continued support. Government research establishments whose tasks are completed or diminishing tend to seek new projects to occupy redundant staff. Also international cooperation can involve a formal obligation which makes withdrawal from untenable projects difficult or simply impossible.

Other arguments why governments are generally not well equipped for direct intervention in industrial innovation refer to the bureaucratic style of governments, their ignorance of economic and technical details, their susceptibility to technical fashions, and their fancies for prestigious projects. Pavitt stresses that governments do not have more information than industry and are not in a position to do a better evaluation of future technical and economic prospects of a project than industry.[19] Jewkes points out that firms or government laboratories, as recipients of government money, have themselves an interest in the matter, so government is unlikely to get unbiased information.

As to the effectiveness of government support to commercial development, Pavitt and Wald draw attention to international comparisons showing no significant correlation between government R&D and national performance in technological innovation. They point to the experience of countries which have managed to have a strong capability in industrial technology with no or very little involvement in large-scale government-subsidized projects, for example, Switzerland, the Netherlands, and Japan.[20]

Eads and Nelson find that large-scale subsidy to commercial technology can also have a degrading effect on government practice beyond the field of R&D. A government, which has a financial stake in a specific commercial venture compounded with political commitments, becomes the lobbyist of its product and is in danger of using its power to create commercial advantages for its favored product at the expense of other products and of overall economic efficiency.[21]

Some authors have even argued that government support of large-scale commercial projects leads to a degradation of industry. Firms dependent on government money will orient their managerial skills to political requirements of lobbying rather than to commercial requirements of marketing and production.[22] Zysman found that government-protected firms retained a centralized and functionally divided organization, whereas unprotected firms modernized their organization and adopted more competitive structures which were decentralized and product-oriented.[23]

Key Questions

The two positions presented earlier have a number of assumptions in common. There is agreement that in some sectors of modern technology the innovation process may involve large cost, long lead times, and high uncertainty about eventual commercial success. This makes them a good starting point for a case study on a nuclear reactor program. Let us now derive from them the questions that may be capable of guiding the following analysis. For this purpose those points have to be singled out where the two positions contradict each other. This leads us to key questions, an answer to which would at the same time corroborate one position and reject the other.

Where did the initiative for the project come from and which organization was the driving force for continuation at crucial decision points? Galbraith would suggest that industry urged government into the project, while Eads and Nelson would suppose that the initiative and pressures came from within the government.

How was the economic need for FBRs assessed and in what terms were the benefits of the project conceived? Both positions assume that the assessment of economic need played a small role in policymaking. Galbraith would insinuate that the participating firms saw the benefits mainly in terms of an underwriting of industrial technology, a contribution to the firms' planning capability, or technological virtuosity. Eads and Nelson would posit that the project was put forward in terms of a technological opportunity to be seized.

Were the costs and risks beyond the reach of the participating firms? Galbraith's position would imply a positive answer. According to Eads and Nelson one would expect that industry declined to finance the project not because of the scale of costs but because of poor commercial prospects. Furthermore, Eads and Nelson would suggest that efforts to make benefits flow earlier increased costs and risks.

Notes

1. National Science Foundation, *Federal Funds for Research and Development, Fiscal Years 1977, 1978, and 1979* (Washington, D.C.: U.S. Government Printing Office, 1979), p. 5.

2. R.R. Nelson, M.J. Peck, and E.D. Kalacheck, *Technology, Economic Growth and Public Policy* (Washington, D.C.: Brookings, 1967).

3. R. Gilpin, "Technological Strategies and National Purpose," *Science* 169 (31 July 1970):441–448.

4. A more extensive discussion may be found in Nelson, Peck, and Kalachek, op. cit., note 2, and J.E. Tilton, *U.S. Energy R&D Policy: The Role of Economics* (Washington, D.C: Resources for the Future, 1974).

5. G. Eads, "US Government Support for Civilian Technology," *Research Policy* 3 (1974):2–16.

6. R.R. Nelson, "The Simple Economics of Basic Scientific Research," *Journal of Political Economy* 67 (1959):297–306.

7. G. Eads and R.R. Nelson, "Governmental Support of Advanced Civilian Technology: Power Reactors and the Supersonic Transport," *Public Policy* 19 (1971):405–427; A Cairncross, "Government and Innovation," in *Uses of Economics,* edited by G. Worswick (London: Blackwell, 1972), pp. 1–20; J. Jewkes, *Government and High Technology,* Institute of Economic Affairs, Occasional Paper 37 (London: Institute of Economic Affairs for the Wincott Foundation, 1972).

8. K. Pavitt and W. Walker, "Government Policies towards Industrial Innovation: A Review," *Research Policy* 5 (1976):11–97, especially pp. 86–88.

9. J.K. Galbraith, *The New Industrial State* (Boston: Houghton Mifflin, 1967).

10. Ibid., p. 33.

11. Ibid., p. 174.

12. Ibid., p. 20.

13. Eads and Nelson, op. cit., note 7; for a revised version see chapter 7 in R.R. Nelson, *The Moon and the Ghetto* (New York: Norton, 1977).

14. T.A. Wise, "IBM's $5,000,000,000 Gamble," *Fortune* (September 1966):118–123, 224–228.

15. R. Miller and D. Sawers, *The Technological Development of Modern Aviation* (London: Routledge and Kegan Paul, 1968); A Phillips, *Technological Change and the Market for Commercial Aircraft* (Lexington, Mass.: D.C. Heath and Company, 1973).

16. R.R. Nelson, "'World Leadership,' the 'Technological Gap' and National Science Policy," *Minerva* 9 (1971):386–399.

17. F.M. Scherer, "Government Research and Development Programs," in *Measuring the Benefits of Government Expenditures,* edited by R. Dorfman (Washington, D.C.: Brookings, 1965), pp. 12–70.

18. Jewkes, op. cit., note 7, pp. 13–14.

19. K. Pavitt, "The Choice of Targets and Instruments for Government Support of Scientific Research," in *The Economics of Industrial Subsidies,* edited by A. Whiting (London: HMSO, 1976), pp. 113–138.

20. K. Pavitt and S. Wald, *The Conditions for Success in Technological Innovation* (Paris: OECD, 1971), p. 126.

21. Eads and Nelson, op. cit. note 7, pp. 424–425.

22. R. Clarke, "Mintech in Retrospect," *Omega* 1 (1973):137–163.

23. J. Zysman, "Between the Market and the State: Dilemmas of French Policy for the Electronics Industry," *Research Policy* 3 (1975):312–336.

2 Nuclear Reactors in West Germany

Historical and Political Context

For various historical and political reasons civilian nuclear power has been developed in West Germany in a decentralized organizational framework which differed from the more centralized structures adopted by the large Western atomic powers.[1] In the United States, Great Britain, and France, government responsibility for nuclear energy has been initially entrusted to a public agency which, in addition to civilian nuclear power, has been responsible also for military applications. After World War II West Germany was excluded by the allies from nuclear research and development (R&D) for military and civilian purposes. In the Paris treaties of 1954, which reestablished national sovereignty and opened the way for civilian power, the country renounced the manufacture of atomic weapons.[2] Hence, civilian nuclear power could develop without being linked to a nuclear weapons program.

Some limits of centralized-government responsibility resulted from the federal structure of government in West Germany. In order for the Federal Government to assume formal responsibility for nuclear power, an amendment of the constitution was necessary which required the approval of the Federal States (Länder). Responsibility for support to scientific research was shared between the Federal Government and the Federal States. This meant that in supporting nuclear science and engineering the states could take action in parallel to the Federal Government.[3]

The christian-democratic governments in power during the 1950s and early 1960s pursued an industrial policy along the principles of the market economy. Thus in the field of nuclear power, design, construction, and operation of power reactors was regarded as an affair of private industry and the utilities, with government lending a supportive and regulatory role.

Government influence was also limited by the decentralized structure of the West German utility industry. Public supply of electricity is provided by several hundred utilities, the ten largest covering more than half of total supply. Most utilities in West Germany are run as mixed enterprises, the Federal States and local communities being usually the dominant shareholders.[4] Apart from price regulation, state and local government exert little influence on business policy. The utilities operate along the lines of private business, as they have to pay an adequate profit to retain the capital of their private stockholders and to attract new private capital for corporate growth. Unlike governments in countries with large state-owned utilities, the West German Federal Government has no formal power to determine directly the procurement policy of utilities.

Early Initiatives

When the allies' ban on nuclear technology was lifted in 1955 with the coming into force of the Paris treaties, work did not have to start from scratch. The German nuclear physics community had survived through theoretical work, and for some experimentation the allies had granted special permission.

A prominent center of nuclear physics in postwar years was the newly founded Institute for Physics of the Max Planck Society under Heisenberg's direction in Göttingen. Heisenberg had worked during World War II on a heavy-water-moderated, natural–uranium–fueled reactor. This work had been terminated by the end of war before a critical reactor had been achieved.[5] After the war, a group of scientists started design studies on a research reactor in the new institute at Göttingen.

As soon as work on nuclear power was permitted, the larger electrical companies, Siemens, Allgemeine Elektricitäts-Gesellschaft (AEG), and Brown, Boveri & Cie (BBC), established design teams for nuclear power plants. Also some manufacturers of heavy-plant equipment such as Krupp, Gutehoffnungshütte Sterkrade (GHH), Maschinenfabrik Augsburg-Nürnberg (MAN), Deutsche Babcock & Wilcox, and Demag showed interest in developing nuclear plants. Krupp joined forces with BBC. Demag combined in 1957 with Atomics International, Division of North American Aviation Inc., in the joint subsidiary Interatom, each company holding 50 percent of a total capital stock of DM9 million ($2.25 million).[6]

Joint ventures were not the only channel by which foreign know-how was accessible to German industry. Some firms, for example Deutsche Babcock & Wilcox, could draw on the technology of foreign associated companies.[7] Others entered in licensing agreements with foreign companies, for instance, GHH in 1960 with General Atomic (then Division of General Dynamics Corporation).[8] Some had long-standing cross-licensing agreements with American firms, which could be extended to the field of nuclear technology (as did AEG with General Electric in 1957) or to specific nuclear components (as did Siemens with Westinghouse in 1957 and 1962).[9] Owing to the late start of reactor development in West Germany, the firms could learn a lot from the experiences in other countries, especially the United States, simply by closely monitoring their activities.

The utilities in general took a wait-and-see stance. They stressed that they were not going to install larger capacities of nuclear plants, unless these became competitive with conventional plants.[10] Nevertheless, they deemed it necessary to collect some experience with small power reactors in order to become knowledgeable buyers. A number of study groups were set up in 1956 and 1957: the Arbeitsgemeinschaft Deutscher Energieversorgungs- unternehmen zur Vorbereitung eines Leistungs-Versuchs-Reaktors e.V. (AVR) by a group of municipal utilities; the Gesellschaft für die Entwick- lung der Atomkraft in Bayern mbH by Bavarian utilities and some firms; the Studiengesellschaft für Kernkraftwerke GmbH (SKW) by utilities in northern Germany and West Berlin; the Arbeitsgemeinschaft Baden-Würt- temberg zum Studium der Errichtung eines Kernkraftwerks (in short Arbeitsgemeinschaft Kernkraftwerk Stuttgart or AKS) by a group of utilities in southwestern Germany.[11]

The Rheinisch-Westfälisches Elektrizitätswerk AG (RWE), the largest West German utility, was the only one to have a study group of its own. As early as 1956, this company decided to order a small experimental power plant of about 10MWe and invited tenders from over ten American and British companies. Seven tenders were received, and a letter of intent was given in March 1957 to a consortium of American Machine and Foundry Atomics, Inc. (United States), Mitchell Engineering Ltd. (Great Britain), and the German firm Siemens for a 15-MWe boiling-water reactor (BWR). By choosing this type of reactor, RWE intended to contribute to a broaden- ing of experience in Europe, as at that time in other European countries only gas-cooled reactors and pressurized-water reactors (PWRs) were being constructed or planned. The contracts were dissolved in mid-1957, however, as plans for a similar reactor in the United States, on whose experience the RWE reactor was to draw, did not materialize.[12]

The AVR association studied nuclear plants with a special view to their possible use in small utilities. Initial studies on Magnox-type reactors were done in collaboration with Deutsche Babcock & Wilcox, which had access to this technology through its foreign associated firms. In 1956 contacts were also established with the consortium BBC/Krupp, which was working on a high-temperature gas-cooled reactor (HTGCR) of the pebble-bed type. A decision was taken in favor of the latter, and a design contract for a 15-MWe plant was placed with BBC/Krupp in April 1957.[13]

By contrast to the wait-and-see stance of the utilities, the attitudes of West German government officials and of the media were deeply influenced by the first United Nations International Conference on the Peaceful Uses of Atomic Energy in August 1955 at Geneva. Though no country had at that time achieved a commercially operating nuclear power plant, the dis- closure of information about foreign activities in nuclear technology hit some of the German representatives and visitors as "a shock." A senior government official in North-Rhine-Westphalia wrote:

Suddenly it became clear to all participants that in the world, totally independently from the atomic bomb, great things had developed in the area of the nuclear sciences. . . . The German public was seized by a deep concern that all these great changes would go past us in Germany and touch would possibly be lost forever.[14]

In the wake of the Geneva Conference, the responsibility for nuclear power in the Federal Government was shifted from the Ministry of Economic Affairs to a new Ministry for Atomic Questions established in October 1955.[15] As an advisory body for the new ministry, the Deutsche Atomkommission (German Atom Commission) was founded in January 1956.[16] It was installed not by a decision of the Atom Minister but by decree of the cabinet, so the minister was obliged to hear the view of the commission on all essential questions.[17] The commission consisted of about twenty-five members, most of them top managers of the large firms in the electric, chemical, metal, banking, and insurance industries, or university professors, preferably physicists. For its recommendations the commission drew on five expert commissions. These set up standing Working Groups which discussed and recommended on the details (see figures B-1 and B-2). Together with its various subdivisions, the commission comprised practically everybody with rank and name from business, universities, and government laboratories who was working in nuclear research and technology or had an interest in nuclear energy, altogether more than 200 individuals. Though the members were appointed personally and not as representatives of their institutions or firms, practically all organizations and firms involved in nuclear energy were represented either in the commission or its subdivisions.[18]

The first Minister for Atomic Questions, Franz Josef Strauss, alloted his ministry the following functions: preparation of a nuclear energy law, preparation of a law on the protection of the population against radioactive materials, drafting of a coordinating program for research, drafting of a program to secure an adequate supply of trained manpower, and international negotiations for the procurement of foreign research reactors and nuclear materials. With regard to the development of power reactors, Strauss announced a rather cautious strategy. One or two small experimental plants were to be constructed within five or seven years. Larger plants of 100 MWe and more were to follow later as a third step.[19]

For the time being the German program was to focus on research reactors. With only one exception, the research reactors were bought directly in the United States and Great Britain or built by domestic firms using foreign designs (see table 2-1). Access to foreign technology had been opened by the American Atoms for Peace program announced by President Eisenhower in 1953. In its wake Great Britain made similar offers. The Ministry for Atomic Energy negotiated agreements with the United States and Great Britain for supply of foreign research reactors and pertinent nuclear materials. It coordinated the various initiatives for research reactors, and gave some financial aid.

A heavy-water research reactor designed at the Max Planck Institute for Physics at Göttingen formed the core of a nuclear research laboratory located near Karlsruhe. The laboratory was constructed jointly by the Federal Government, the Land Baden-Württemberg, and industry. The state government of North-Rhine-Westphalia erected its own nuclear reseach laboratory, the Kernforschungsanlage Jülich, as a common research facility for all the universities and engineering schools (Technische Hochschulen) in this state. Two British research reactors were bought for this laboratory. They were financed by the Land North-Rhine-Westphalia, together with a substantial grant to the investment costs from the Federal Government.

A laboratory for ship reactors, the Gesellschaft für Kernenergieverwertung in Schiffbau und Schiffahrt mbH, was founded in April 1956. It was a joint venture of four Länder on the North Sea and Baltic Sea coasts and a number of firms, among them some shipyards. For its use an American research reactor was purchased, the costs of which were shared among Länder authorities, Federal Government, and industry. In Berlin, an American reseach reactor formed the focus of a laboratory founded by the senate of Berlin under the name Hahn-Meitner-Institut für Kernforschung. Its investment costs were financed jointly by the Land Berlin and the Federal Government. Two American research reactors were bought for the Technical University at Munich and the University of Frankfurt. Siemens and AEG each constructed a research reactor with their own funds and for their own purposes, using a design of the Argonne National Laboratory in the United States. A reactor of the same type was ordered by the Karlsruhe laboratory from German firms.

In general, government support of the early research reactors conformed to the federal structure of West Germany, since the Federal Government assumed only a supportive role, though in some cases its financial contribution was substantial. In later years the financial stake of the Federal Government in all the nuclear research laboratories continually increased, while industry's financial contributions ceased with completion of the first research reactors. By the early 1970s all laboratories were financed 90 percent by the Federal Government and 10 percent by the Länder governments.[20]

Eltville Program

In 1957 the German Atom Commission considered a way in which the efforts of different manufacturers and utilities on power reactors could be coordinated and the kind of government support to be given. The task of

Table 2–1
Research Reactors Ordered until 1960

Designation, Site, Commissioning Date	Type, Output (thermal kilowatts)	Supplier	Operator	Source of Funds
FRM München-Garching October 1957	Open swimming pool 1,000 kW	American Machine & Foundry, MAN	Laboratorium für Technische Physik, Technische Hochschule München	Land Bayern, Fed. Government
FRF Frankfurt am Main January 1958	Homogeneous, water boiler 50 kW	Atomics International, AEG, BBC, Mannesmann, Siemens	Institut für Kernphysik, Universität Frankfurt	Hoechst (reactor) Fed. Government (equipment) City of Frankfurt (building)
BER Berlin July 1958	Homogeneous, water boiler 50 kW	Atomics International, AEG, Borsig, Pintsch-Bamag, Siemens	Hahn-Meitner-Institut für Kernforschung	Land Berlin, Fed. Government
FRG 1/2 Hamburg Geesthacht October 1958/March 1963	Open swimming pool with 2 cores 5,000 kW per core	Babcock & Wilcox (United States), Deutsche Bacock & Wilcox	Gesellschaft für Kernenergieverwertung in Schiffbau und Schiffahrt mbH	Fed. Government (33%), Länder Bremen, Hamburg, Niedersachsen, Schleswig-Holstein (33%), industry (33%)
FR2 Karlsruhe March 1961 (44 MW: 1966)	Closed swimming pool 12,000 kW (later 44,000 kW)	Domestic firms; design by the Karlsruhe research center	Kernreaktor Bau- und Betriebsgesellschaft mbH	Fed. Government (30%), Land Baden-Württemberg (20%), industry (50%)

FRJ 1 MERLIN Jülich February 1962	Open swimming pool 5,000 kW	A.E.I.-John Thompson Nuclear Energy Company, Ltd. (Great Britain), AEG, Rheinstahl	Kernforschungsanlage Jülich	Land Nordrhein-Westfalen, Fed. Government
FRJ 2 DIDO Jülich November 1962	Closed swimming pool 10,000 kW	AEG, Rheinstahl, plans by Head Wrightson Processes Ltd. (Great Britain)	Kernforschungsanlage Jülich	Land Nordrhein-Westfalen, Fed. Government
SAR München-Garching June 1959	Argonaut 1 kW (short time: 10)	Siemens, plans by Argonne National Laboratory (United States)	Siemens	Siemens
PR 10 Grosswelzheim January 1961	Argonaut 0.1 kW	AEG, plans by Argonne National Laboratory	AEG	AEG
ARGONAUT (later STARK) Karlsruhe January 1963	Argonaut (later modified) 0.007 kW	Siemens, Lurgi, Pintsch-Bamag, plans by Argonne National Laboratory	Gesellschaft für Kernforschung mbH	Fed. Government (75%) Land Baden-Württemberg (25%)

Sources: L. Brandt, "Die gemeinsamen Atomforschungsanlagen in Nordrhein-Westfalen," *Atomwirtschaft* 2 (1957):247–252; A. Weber, "Forschungsreaktoren," *Atomwirtschaft* 3 (1958):302–307; "Argonaut Karlsruhe kritisch," *Atomwirtschaft* 8 (1963):138; R. Gerwin, *Atomenergie in Deutschland* (Düsseldorf and Vienna: Econ, 1964), pp. 154–159; J. Sobotta, ed., *State, Science and Economy as Partners* (Berlin and Vienna: Koska, 1967), pp. 166–167.

drafting a program in detail was given to the Working Group II-III/1 on Nuclear Reactors (henceforth Working Group). Membership of the group comprised two individuals from firms interested in the manufacture of reactors, one from a utility, three from firms in the chemical industry, and three scientists of professorial rank from universities or research centers.

The Eltville Program, named after the place of one of the Working Group's meetings, was approved by the Atom Commission on 9 December 1957. It envisaged design and construction of five nuclear power plants of different types at ratings of 100 MWe, each by a different domestic supplier (see table 2-2). The aim was to install a total capacity of 500 MWe until 1965. In addition to this 500 Megawatt Program, plants up to a total capacity of 1,000 MWe were to be bought directly from foreign suppliers provided they were economic.[21]

Coordination of the domestic manufacturers within the 500 Megawatt Program was not difficult. Four firms or consortia were interested in reactor construction, and each proposed for the program a different type of reactor: Deutsche Babcock & Wilcox a Magnox reactor, Siemens a heavy-water reactor, AEG a light-water reactor (LWR), and BBC/Krupp a high-temperature reactor. Provisions were even made for an additional firm which did not yet exist: Interatom, the joint subsidiary of Demag and Atomics International (United States), founded in December 1957. In line with the activities of its American parent company, Interatom began to work on an organic-cooled reactor.

Specifications for reactor types were deliberately broad and elastic. For instance, the type of reactor to be developed by AEG did not specify whether this was an LWR of boiling-water type or of pressurized-water type. In general, the 500 Megawatt Program was biased toward reactors fueled by natural uranium, since continuous supply of enriched uranium in adequate amounts was not regarded as secure at that time.

Not all reactor activities of the firms were associated with the 500 Megawatt Program. Siemens, for example, had initially worked on three designs: the gas-graphite reactor, the heavy-water-moderated and heavy-water-cooled reactor using a pressure vessel, and the heavy-water-moderated gas-cooled reactor using pressure tubes. The firm then concentrated its self-financed efforts on the heavy-water-moderated and heavy-water-cooled reactor with a pressure vessel, which appeared to be the most promising one. The heavy-water gas-cooled reactor was continued under the government program, while the gas-graphite reactor was dropped.[22]

The 500 Megawatt Program envisaged that construction costs of the five reactors could be financed by the utilities, with the aid of special tax reductions and government loans and subsidies. Nuclear plants were expected to cost about three times more than the same capacity conventional plant. The program assumed that the utilities would provide from their own sources an amount equivalent to the costs of conventional capacity, which would cover one-third of total investment costs. Another third would be financed through tax concessions, and the rest through government loans or subsidies which would be repayable if the plants made a profit. Design work was to be funded one-third by the utilities and two-thirds through government loans which, if a plant were to be constructed, would be considered as part of the government loan to the investment costs. In July 1958 it was decided that the government should also share eventual operating losses of the plants.[23]

While the 500 Megawatt Program largely reflected the concurrent views of supplier industry and government, the utilities opposed it. They thought 100-MWe plants involved too large a risk, and advised the design and construction of 15-MWe plants instead.[24] Whereas the government program assumed that the question as to which group of utilities was to order which type of reactor would be but a formality, the utilities rejected this limitation of their choice. They stressed that economic aspects rather than the requirements of the government program would determine their choice of reactor, if they built a plant at all. Within the Atom Commission, representatives of reactor suppliers and individuals from research laboratories and government agencies stressed that the primary aim of reactor development was to establish an internationally competitive domestic industry, and that this aim should not be hindered by what they regarded as a premature and overstressed bias toward economic aspects. Rather than reduce the rating of the plants, this group would have liked to see the program implemented without the cooperation of the utilities, even if this conflicted with the official industrial policy at that time.[25]

Underlying this conflict were other differences. The utilities were cautious in their appreciation of the role of nuclear power in the near future. They considered any forecasts of future nuclear capacities as irresponsible.[26] They saw no economic need for accelerating the development of nuclear power. For them it seemed both possible and desirable to assure an adequate supply of electricity by developing domestic conventional sources and by rational energy use rather than by an early deployment of uneconomic and technically immature nuclear plants. Influential individuals in utilities deemed it desirable to delay rather than accelerate the commercial introduction of nuclear power in order to have technologically sound plants when finally needed. Heinrich Schöller, RWE's board member in charge of plant construction, wrote: "This way we avoid wasting billions [of deutschmarks] which because of early obsolescence or technical failures of such plants would have to be regarded as misinvestments and which, if used for increasing the production of domestic fuels, would yield a better return."[27]

Furthermore, the utilities conceded only a limited role to government

Table 2-2
The 500 Megawatt Program of 1957

Firm	Type of Reactor (fuel, moderator, and coolant)	Association of Utilities Placing Design Contract (date of foundation)	Outcome	Expenditure by Government and Utilities (in DM million)
Deutsche Babcock & Wilcox	Advanced reactor type Calder Hall (natural uranium, graphite, gas)	SKW Studiengesellschaft für Kernkraftwerke (1957)	Design contract placed 24 March 1959; no reactor of this type constructed	*Design: 6.0* Fed. Government: 4.0 Utilities: 2.0
Siemens	Heavy-water reactor (natural uranium, heavy water, carbon dioxide, with pressure tubes	Gesellschaft für die Entwicklung der Atomkraft in Bayern mbH (1957)	Design contract placed 1 March 1959; KKN-reactor constructed 1966–1974; shutdown before achieving full power	*Design: 8.7* Fed. Government: 5.8 State Government: 0.8 Utilities: 2.1 *Construction: 225.9* Fed. Government: 130.9 Industry: 10.0 Loans: 85.0
BBC/Krupp	High-temperature reactor (enriched uranium, graphite, gas)	AVR Arbeitsgemeinschaft Versuchs-Reaktor GmbH (1959), successor to Arbeitsgemeinschaft deutscher Energiever-sorgungsunternehmen zur Vorbereitung der Errichtung eines Leistungs-Versuchs-Reaktors e.V. (1956)	No design contract placed for 100-MWe plant; contract for 15-MWe plant on 25 April 1957; no reactor of 100 MWe constructed; 13-MWe AVR plant constructed 1960–1969	(AVR design and construction was originally not part of the Eltville Program)

AEG	Light-water reactor (enriched uranium, light water, light water)	SKW Studiengesellschaft für Kernkraftwerke (1957)	Design contract placed 24 March 1959; no reactor of 100 MWe constructed; design used for the Lingen plant (see table 2–4)	*Design: 6.0* Fed. Government: 4.0 Utilities: 2.0
Interatom	Organic-cooled reactor (natural or enriched uranium, heavy water or organic, organic)	Kernkraftwerk Baden-Württemberg Planungsgesellschaft mbH (1960), successor to AKS Arbeitsgemeinschaft Baden-Württemberg zum Studium der Errichtung eines Kernkraftwerks (1957)	No design contract placed for 100-MWe plant; contract for 150-MWe and 15-MWe plants in spring 1961; no reactor of this type ever constructed	*Design: 4.0* Fed. Government: 2.0 Utilities: 2.0

Sources: "Memorandum der Deutschen Atomkommission zu technischen, wirtschaftlichen und finanziellen Fragen des Atomprogramms," 9 December 1957 (unpublished); G. Leichtle, "Leistungsreaktoren," *Atomwirtschaft* 3 (1958):307–310; J. Pretsch, "Leistungsreaktoren in Deutschland," *Atomwirtschaft* 4 (1959):507–508; J. Pretsch and H. Kühne, "Bau und Entwicklung von Reaktoren für Kraftwerke und Schiffsantrieb," in *Taschenbuch für Atomfragen 1960/61*, edited by W. Cartellieri et al. (Bonn: Festland Verlag, 1960), pp. 109–122; "Leistungs- und Forschungsprogramm . . . ," *Atom und Strom* (November 1961):92; financial data for the KKN reactor are from *Bundeshaushaltsplan 1972*, Comments on Kap. 3105 Tit. 89310 and from an unpublished document of the Federal Ministry for Research and Technology.

support. In Schöller's view, government support of nuclear research and tax concessions for small experimental plants were sufficient:

> They [the limits of government support] are in my opinion . . . there, where electricity generation is ventured for the first time, . . . that is, the construction of the first experimental plants should be left to individual business enterprises or to voluntary associations of such enterprises, whereby tax concessions may possibly become necessary in specific cases (e. g. through early depreciation of economic plants) . . . Only their [the enterprises'] own risk, which nevertheless will remain associated with their projects, secures a transition from experiment to economic practice with minimum possible misinvestment.[28]

Finally, the role of foreign technology was a matter of dispute. From the beginnings of nuclear power, the utilities based their cost evaluations on the technology available on the international market. This implied international invitations for tenders, and a willingness to purchase foreign plants if these were superior to domestic designs. Contrary to this policy, supplier firms, the ministry, and research laboratories demanded that domestic technology have a chance first. West German firms felt discriminated against by RWE's international invitations for tenders, despite the utility's assurance that domestic suppliers would participate in the construction.[29] The tensions were accentuated by the utilities' attitude toward supply of enriched uranium. Whereas most German manufacturers regarded supply of enriched uranium to be problematic and hence focused on natural uranium reactors, RWE held that this problem should be disregarded for small experimental plants.[30]

By incorporating in the Eltville Program a supplementary program for 1,000 MWe of foreign plants in addition to 500 MWe of domestic plants, the German Atom Commission circumvented a clear recommendation on the role of foreign technology versus domestic technology. This supplement had been included in the later stages of the formulation of the program as a response to plans made at the European level.[31] The idea of importing American power reactors was an essential element in the report of the "three wise men,"[32] which was a major step in the formation of the European Atomic Energy Community (Euratom) in January 1958 by Belgium, France, West Germany, Italy, Luxemburg, and the Netherlands.[33] The supplement of 1,000 MWe of foreign plants in the Eltville Program was thus a concession to both the political aim of European integration and the technological strategy of the West German utilities. The Eltville Program laid down that the foreign plants were to be ordered only if these were economically competitive with conventional power. This implied that they would need no government support and would not draw government funds away from the 500 Megawatt Program.

Implementation of the 500 Megawatt Program suffered seriously from the utilities' resistance. About two years passed before the first three design contracts were agreed on. Meanwhile the AVR association as well as RWE went on with their plans independently of the government program. When in April 1957 (that is practically during the time when the Eltville Program was devised) the AVR association placed a design contract with BBC/Krupp for a 15-MWe pebble-bed high-temperature gas-cooled reactor (HTGCR), government officials stressed that this reactor was not considered part of the government program,[34] and the design contract was given without government support. For construction of the plant, however, the utility group, refounded in February 1959 as Arbeitsgemeinschaft Versuchs-Reaktor-GmbH (AVR), did receive support from the Federal Government.[35] A contract was signed in August 1959. The AVR reactor, with a net rating of 13 MWe, was constructed at the site of the Jülich research center. Because of technical problems the plant did not supply its first electricity until December 1967. The final costs amounted to DM85.3 million, of which the AVR group paid DM20 million, the Federal Government DM56.2 million, and the supplier DM9.1 million.[36]

After RWE's first project with American Machine & Foundry, Mitchell Engineering, and Siemens had been canceled, the utility renewed its invitation for tenders. Out of four tenders received, an order was placed in June 1958 with AEG for a 15-MWe BWR, the nuclear-steam-supply system being subcontracted to General Electric Company (GE; United States).[37] Bayernwerk AG joined in this venture with a 20-percent share. The plant was constructed at Kahl/Main, hence named Versuchsatomkraftwerk Kahl (VAK). It was completed in a fairly short time and produced in June 1961 the first nuclear electricity in West Germany. The costs of the plant amounted to DM48.7 million (including DM12.7 million for the first core). They were financed by the two utilities.[38] The financial contribution of the government was limited to a DM0.15 million subsidy for licensing costs.[39]

Schöller of RWE stated that the VAK plant was to demonstrate that a development strategy is possible that is better than the program of the Atom Commission:

> The state should occupy itself with research and development; this has been always its primordial task. When we build the nuclear experimental plant Kahl, we do something that actually is not our task, but the task of machine industry. And if I want to be quite frank, my decision to order this plant was determined by the reflection that, if already the state wants to make follies with premature construction of plants, then we better want to make such follies ourselves in order to keep them under control.[40]

In the summer of 1957 RWE invited tenders for plants of more than 100 MWe from British firms, and later also from American and French firms.[41]

The utility's annual report of 1957–1958 emphasized that this was a strategy competing with the 500 Megawatt Program: "With them [the tenders] we have already a valuable basis like that to be gained according to the German 500 MW experimental program through design contracts. But that would be available only in several years."[42]

Although the policy of Euratom had influenced the formulation of the Eltville Program, it contributed little to the undermining of its implementation. A joint Euratom–United States power-reactor program, concluded in May 1958, proposed to import American plants with a total capacity of 1,000 MWe. This meant approximately six reactors about 150 MWe. The reactor types included in this program were the PWR, the BWR, and, after additional negotiations, the organic-moderated reactor. The United States provided long-term credits for construction, and a financial contribution of $50 million to a common research program related to these reactor types. They were also prepared to supply enriched uranium at generous terms, to reprocess spent fuel, and to buy back the plutonium produced in these reactors.

Proposals for nuclear plants under this scheme were invited in April 1959. By the deadline of 20 October 1959, the AKS association and the Berliner Kraft und Licht AG (Bewag) submitted conditional proposals: AKS for a 150-MWe plant with an organic-moderated reactor to be supplied by Atomics International (United States) and Interatom, Bewag for a reactor of unspecified design. However, AKS decided in January 1960 not to build, and Bewag later postponed the project indefinitely. Apart from technical problems, AKS considered the venture both too costly and too risky. Even with subsidies as provided through the Euratom–United States scheme, the plant would have had electricity costs unjustifiably above those of conventional plants, and even a guarantee of the West German government to cover half of the operating losses up to a total of DM100 million did not provide enough insurance for AKS to risk its shareholders' money.[43]

For the West German reactor firms the final result of the Eltville Program (see table 2–2) was as follows. Siemens received a design contract worth DM8.7 million for its heavy-water gas-cooled reactor in March 1959. The design was completed in 1963. After years of negotiations an order for this reactor was given in 1966, and it was constructed at Niederaichbach under the name Kernkraftwerk Niederaichbach (KKN) at a cost of DM225.9 million. However, the financial scheme for this reactor was totally different from that envisaged in the Eltville Program: in addition to tax concessions and guarantees for bank loans of DM35 million, the Federal Government gave a direct subsidy of DM130.9 million. The state Bavaria provided a DM10 million loan free of interest and guaranteed further bank loans of DM40 million. Siemens as supplier contributed DM10 million. The utilities refrained from investing their own money. By the time construction was completed, this technology was obsolete. The plant was shut down in July 1974 before having reached full output.[44]

Independently of the Eltville Program and with its own money, Siemens developed the heavy-water pressure-vessel reactor. At the end of 1959 a detailed design for a 100-MWe plant was completed. The plant was constructed with an output of 50 MWe at the site of the Karlsruhe laboratory, mainly with government money. In addition to generation of electricity, this reactor was to produce plutonium and to provide facilities for irradiation tests, and hence was named Mehrzweckforschungsreaktor (MZFR), that is, multiple-purpose research reactor. The irradiation facilities were included as a research component to circumvent the reservations of the Ministry of Finance against subsidizing commercial reactors, but they were never used. The plutonium produced in this reactor was deemed useful for R&D on reactors fueled by plutonium together with natural uranium, which would avoid reliance on foreign enriched uranium. The order was given in December 1961, and the first electricity was produced in March 1966.[45] The total costs of DM153.5 million were financed by the Federal Government (DM123.5 million), the Land Government of Baden-Württemberg (DM20 million), and two utilities operating the plant (DM10 million).[46] After 1960, when supply of foreign enriched uranium appeared possible on a larger scale, Siemens gave priority to the light-water version of the pressure-vessel reactor. This design appeared less complicated and less costly to build. Also, the size of the pressure vessel of the heavy-water version was seen to set technical limits to scaling up to larger unit sizes.[47]

AEG was given in March 1959 a design contract worth DM6 million for a BWR with steam superheating. As mentioned earlier, the company received in June 1958, independently of the Eltville Program, the order for the 15-MWe VAK plant from RWE and Bayernwerk. A BWR plant of 100 MWe as envisaged in the Eltville Program was never built. The studies performed by AEG under this program were used later for a plant with a 160-MWe nuclar system and fossil steam-superheating, which was ordered in 1964 for a site near Lingen.[48]

Babcock & Wilcox received in March 1959 a design contract worth DM6 million for its "advanced reactor, type Calder Hall." After a half year's work the firm realized that, despite some possible technical improvements, this reactor would remain uneconomic, and the project was changed to an advanced gas-cooled reactor (AGR) with enriched uranium as fuel instead of natural uranium.[49] Even with this new specification the design contract was not followed up by an order. A reactor of Magnox or advanced gas-cooled type was never constructed in West Germany.

BBC/Krupp received no design contract for a 100-MWe HTGCR. As said earlier, the AVR association placed instead, independently of the Eltville Program, a design contract for a 15-MWe plant in April 1957. Against the early intentions of the Eltville Program, this plant was constructed with substantial government aid.

For the organic-cooled reactor of Interatom a design contract worth DM4 million was given as late as spring 1961, probably with a view to the second invitation for proposals in the Euratom—United States power-reactor program. Against the intentions of the Eltville Program, the design contract was for two reactors of 150 MWe and 15 MWe instead of a 100-MWe reactor. No reactor of this type was ever constructed in West Germany.

From a formal view of planning, the Eltville Program failed miserably. Nevertheless, it may have been wise to implement so little. Three of the five reactor types (the Magnox reactor, the heavy-water gas-cooled reactor, and the organic-cooled reactor) were shown by later experience to be bad bets against the LWR. The HTGCR turned out to be a possible competitor to the LWR only in the long run and with much additional development required. The LWR, after all, was to find its way without substantial government support.

Out of the expenditure of DM600 million to DM800 million originally envisaged under this program, only DM250.6 million were spent (funds by manufacturers not included). Of this sum DM24.7 million were for design studies under the first phase of the program, and DM225.9 million for construction of the KKN plant. The dismal experience with this plant suggests that without its construction and with expenditure of DM225.9 million less, the benefit of the program would not have been much impaired. For those reactor types which were later selected for commercial use, only DM6 million of government and utility funds were spent under this program. These were for the BWR, whereas the PWR did not feature at all in this program.

Program for Advanced Reactors

At the end of 1959, only two industrial reactor teams had received orders for small power plants. It was obvious that neither the Eltville Program nor the Euratom–United States program would soon result in further orders. In this situation the Atom Ministry invited industry to submit proposals for a program of small experimental plants of advanced reactor types. The designs were to be finished after three years and the plants were to be constructed in three more years. To facilitate financial support by the Federal Government, the program was framed as experimental. Commercial aspects were to be disregarded in favor of new technical ideas and, unlike in the Eltville Program, the utilities' collaboration was not sought. A senior civil servant in the ministry later characterized this program as "an emergency assistance measure or a stopgap program for industry." He described the ministry's perception of the situation as follows:

It was a time of stagnation; reactor designs developed into title-one designs but there was nobody to place an order for construction. It was also a time of uneasiness for the reactor study groups of industry which to set up had been troublesome and expensive, too, and which were now on the edge of breaking up or even migrating.[50]

The firms' proposals were received in March 1960 and assembled into what was later called the Program for Advanced Reactors (see table 2–3). The various committees of the German Atom Commission gave their approval in the first half of 1960.[51] The program envisaged design and construction of a small experimental reactor for each firm submitting a proposal. AEG suggested a BWR with nuclear steam superheating. Work on such a design was at that time also done in the United States. Deutsche Babcock & Wilcox proposed an AGR, continuing its work under the 500 Megawatt Program. A similar design was developed in Britain at that time.

The proposal of Brown Boveri/Krupp was for a quasi-homogeneous HTGCR with a gas turbine. From the reactor ordered by the AVR association in 1959, this design differed in the use of a gas turbine and in the fuel element. In the initial AVR design the fuel had the form of a pellet which was inserted in graphite balls of 6-cm diameter. According to the new design, the fuel was to be distributed evenly in the graphite balls.[52]

Interatom originally suggested an organic-cooled reactor, but during the preparation of the program changed to a sodium-cooled zirconium-hybride-moderated reactor. Work on this reactor type was done at this time also in the United States.

As in the Eltville Program, the coordination achieved in the Program for Advanced Reactors was not difficult. Apart from Siemens which was preparing for construction of the government-financed 50-MWe MZFR, every firm interested could participate in the program and pursue the type of reactor it wished. Giving support to every firm interested in nuclear reactors seems to have been at that time a general policy of the Atom Ministry. For instance, when Maschinenfabrik Augsburg-Nürnberg (MAN) in 1960 asked for government support for design work on a 5-MWe PWR of horizontal type, the Working Group refused approval on the ground that this design was licensed abroad and not a domestic development. The ministry did not follow the Working Group's advice and financed 50 percent of MAN's design work on this reactor in subsequent years.[53]

Contrary to the Eltville Program, the design phase of the Program for Advanced Reactors could begin without delay, as government funds were appropriated directly to manufacturing industry without the utilities' cooperation. Out of the four reactors only two were later built: the BWR with integral steam superheating of AEG, later called Heissdampfreaktor (HDR) and the zirconium-hydride-moderated sodium-cooled reactor of Interatom, later called Kompakte Natriumgekühlte Kernreaktoranlage (KNK). Construction of these two reactors was recommended by various bodies of

Table 2–3
Program for Advanced Reactors of 1960

Firm/Consortium	Type of Reactor (rating as planned in 1960)	Outcome (name of reactor rating, construction schedule)	Design (m DM)	Expenditure	
				Experimental R&D (m DM)	Construction (m DM)
AEG	Boiling-water reactor with nuclear steam superheating, 25 MWe	HDR reactor, 23 MWe, 1965–1969, (shut down 1971)	FG: 16.0 S: 4.0	FG: 25.3 S: 2.7 Others: 1.3	FG: 81.0
Deutsche Babcock & Wilcox	Advanced gas-cooled reactor, 10–15 MWe	No reactor constructed	FG: 12.9 S: 3.2	—	—
BBC/Krupp	Homogeneous high-temperature gas-cooled reactor with gas turbine, 5–10 MWe	No reactor constructed; studies used for AVR reactor	FG: 9.5 S: 2.4	—	—
Interatom	Zirconium-moderated sodium-cooled reactor, 10 MWe	KNK reactor, 18 MWe, 1966–1971[a]	FG: 18.0 S: 4.5	FG: 11.5	FG: 128.8 U: 4.0 (S: 10–20)

Sources: H. Kühne, "Betrieb, Bau und Entwicklung von Reaktoren in der Bundesrepublik Deutschland," in *Taschenbuch für Atomfragen 1964*, edited by W. Cartellieri et al. (Bonn: Festland Verlag, 1964), pp. 88–100; *Bundeshaushaltsplan 1961*, Comments on Kap. A 3102, Tit. 891; Financial data for KNK reactor are from *Bundeshaushaltsplan 1972*, Comments on Kap. 3105 Tit. 89310, data for other items from an unpublished document of the Federal Ministry for Research and Technology.

[a] Construction start and first criticality.

Abbreviations: FG: Federal Government; S: supplier; U: utilities.

the German Atom Commission as early as 1963,[54] but did not start before 1965 and 1966, respectively. The HDR reactor cost DM81 million and was financed totally with government money. The costs of the KNK reactor amounted to DM128.8 million, of which the utility operating the plant contributed DM4 million, and the Federal Government paid the rest. Due to a fixed-price contract, the supplier Interatom is reported to have incurred cost-overruns of between DM10 million and DM20 million.[55]

As in the case of the Eltville Program, it was probably wise not to implement the whole program. The AGR was never accepted by German utilities as a competitor to LWRs. As for the HTGCR, both fuel concepts, pellet and homogeneous, ran into technical difficulties and were abandoned in 1963 in favor of fuel in the form of coated particles as developed in the United States.[56] Apart from the gas turbine, the design proposed by Brown Boveri/Krupp for the Program for Advanced Reactors would then have only duplicated the AVR reactor. The HDR and the KNK reactors were never followed up by orders for larger plants. Shortly after operation startup, the HDR plant suffered a failure in some fuel elements. At that time, AEG had no further interest in this technology, so the plant was shut down.[57] When the design of the KNK plant was completed around 1963, sodium-cooled zirconium-moderated reactors no longer appeared commercially attractive.[58] However, the KNK plant was constructed for the West German fast breeder program, as it was to yield experience with sodium-cooling and as it could be later converted into a test facility for fast breeder fuel elements.[59]

Light-Water Reactor Demonstration Plants

By 1961, only one small nuclear plant was in operation (the 15-MWe VAK) and two small plants were under construction (the 13-MWe AVR and the 50-MWe MZFR). This did not seem to be an impressive record of six years' work by the Atom Ministry. Hence the ministry determined that German plants would materialize, when a second invitation under the joint Euratom–United States program was issued in July 1961.[60]

RWE and Bayernwerk had been studying during 1957–1962 eleven tenders for a larger nuclear plant from six firms or consortia, including American, British, and French manufacturers. The studies had come down by 1961 to a choice between a 300-MWe gas-graphite plant of a consortium of English Electric (Great Britain), Babcock & Wilcox (Great Britain), Siemens, and Deutsche Babcock & Wilcox, and a 240-MWe boiling-water plant of a consortium of GE (United States) and AEG.[61] Siemens had combined for this bid with English Electric and Babcock & Wilcox, because its development work on the pressurized LWR had not yet proceeded to the point where a bid could be ventured.[62]

RWE and Bayernwerk AG chose the GE/AEG bid as the successful one, because the BWR of American design was, at the terms of the joint Euratom–United States program, more economic than the gas-graphite reactor of British design. The plant was constructed at a site near Gundremmingen (see table 2–4). Owner and operator of the plant is the Kernkraftwerk RWE-Bayernwerk GmbH (KRB), a joint venture of RWE and Bayernwerk with a capital of DM100 million, with 75 percent of the capital contributed by RWE and 25 percent by Bayernwerk. Euratom gave DM32 million to the investment costs of the plant and granted KRB the status of a common European enterprise, which conferred some privileges with regard to taxes and customs. DM172 million were financed through American and German bank loans, which were guaranteed by the Federal Government. The U.S. Atomic Energy Commission (AEC) gave a guarantee to provide reprocessing services and to buy the plutonium produced in this reactor. The Länder governments of Bavaria and North-Rhine-Westphalia provided special allowances for depreciation to the parent companies of KRB. The Federal Government guaranteed operating losses up to DM 100 million, of which DM18 million were actually claimed up to 1972. Under this scheme of support the plant could operate roughly competitive to a coal plant at the same site, which was far from the West German coal regions.

In the various bodies of the Atom Commission the Gundremmingen order accentuated dissent about the role of foreign technology. Individuals from government laboratories and from domestic manufacturers opposed government support for this plant. They argued that an import of foreign plant did not accord with the German nuclear program which ought to support domestic development. Despite this resistance, government support was finally approved by the commission and its subdivisions. They expressed the expectation, however, that government support would be available also for the construction of nuclear plants of domestic design.[63]

The Working Group decided that the issues should be cleared in a special meeting with representatives of the major West German utilities. This meeting took place on 11 February 1963. The minutes reveal a fundamental dissent on all major questions of a development strategy for nuclear power.[64] Opinions differed as to whether natural-uranium reactors (including reactors with plutonium recycling) should be given preference to enriched-uranium reactors, so as to avoid dependence on foreign supply of enriched uranium. No consensus existed as to whether future development work was to focus on proven reactors such as the LWR or to proceed directly to advanced types of reactors such as the HTGCR or the fast breeder reactor (FBR). Diverging opinions were expressed as to whether domestic designs should be favored against foreign technology. The only point of agreement concerned state support to nuclear power which all sides regarded as a necessity.

In 1964 and 1965, orders for two more LWR demonstration plants were placed (see table 2-4). A 240-MWe plant with a BWR and fossil steam superheating was ordered for construction at Lingen. The owner and operator was Kernkraftwerk Lingen GmbH, founded jointly by the utility Vereinigte Elektrizitätswerke Westfalen AG (VEW), the supplier AEG, and a consortium of banks. AEG supplied the plant independently from assistance by GE. The design of the reactor was a continuation of AEG's design work for a 100-MWe BWR under the 500 Megawatt Program.

A third LWR demonstration plant was constructed in Obrigheim, for which a bid by Siemens for a PWR was chosen against a bid by AEG for a BWR. The plant was supplied at a guaranteed power of 283 MWe, and was stretched in 1969 to 328 MWe. It was mainly based on Siemens's own development work, though for some components Westinghouse licenses were drawn upon. The owner/operator was Kernkraftwerk Obrigheim GmbH (KWO), a successor to the AKS association.

Government support for the Lingen and Obrigheim plants followed roughly the same scheme as for the Gundremmingen plant. The investment costs were covered by owners' stock and by loans guaranteed by the government. The Federal Government contributed subsidies of DM40 million for each of the plants, which were earmarked for the first core of the plants and for the research and development programs associated with the plants. The R&D grant was in fact a subsidy for the construction costs, since it was given on condition that the supplier firms reduce their bids by the same amount. For each plant the government gave a guarantee for operating losses up to DM100 million, of which around DM20 million were claimed up to 1972 for the Lingen plant, and around DM10 million for the Obrigheim plant. Reductions for taxes and customs were obtained by receiving the status of a common European enterprise. Furthermore, the financial burden of the parent companies of the owner-operator firms was reduced by special allowances for depreciation.

Commercial Phase

In 1967, two nuclear power plants were ordered which did not need government subsidies for construction and operation. They can be regarded as the beginning of the commercial phase of nuclear power in West Germany. AEG received an order for a 640-MWe plant with a BWR from Preussische Elektrizitäts-Aktiengesellschaft, and Siemens for a 630-MWe plant with a PWR from Nordwestdeutsche Kraftwerke AG and Hamburgische Elektricitäts-Werke. From 1969 to 1975 there was a series of commercial orders, all of which were for LWR plants. Prodded by the utilities, mainly by RWE, whose Biblis plant was at the time of operation startup the largest nuclear

Table 2–4
Light-Water Reactor Demonstration Plants

	Gundremmingen	Lingen	Obrigheim
Type of reactor	Boiling-water reactor	Boiling-water reactor with fossil steam superheating	Pressurized-water reactor
Rating	237 MWe	240/256 MWe[a] of which nuclear 160 MWe fossil 80/96 MWe[a]	283/328 MWe[a]
Owner/operator (date founded)	Kernkraftwerk RWE-Bayernwerk GmbH (24 July 1962)	Kernkraftwerk Lingen GmbH (3 March 1964)	Kernkraftwerk Obrigheim GmbH (25 November 1964)
Shareholders of owner/operator company	RWE (75%), Bayernwerk AG (25%)	Vereinigte Elektrizitätswerke Westfalen AG (34,375%), AEG (34,375%), a consortium of banks (31,250%)[b]	Energie-Versorgung Schwaben AG (35%), Badenwerk AG (28%), Technische Werke der Stadt Stuttgart AG (14%), Neckarwerke Elektrizitätsversorgungs-AG (10%), and nine other utilities
Supplier	AEG, International General Electric Operations S.A., Hochtief AG	AEG	Siemens
Date of order	10 November 1962	19 June 1964	12 March 1965
First criticality	14 August 1966	31 January 1968	22 September 1968
Start of operation[c]	23 December 1966	20 July 1968	15 March 1969
Shut down	13 January 1977	5 January 1977	in operation

Capital costs by source of funds	(DM million)	(DM million)	(DM million)
Utilities	100	80	100
Government[d]	32	40	40
Loans	172	150	191
Total	304	270	331

Operation losses until 1972 by source of funds	(DM million)	(DM million)	(DM million)
Government	18	19	10
Utility	—	2	—
Total	18	21	10

Sources: G. Schuster, "Deutsches Reaktorprogramm," in *Taschenbuch für Atomfragen 1968*, edited by W. Cartellieri et al. (Bonn: Festland Verlag, 1968), pp. 147–171; "KRB Kernkraftwerk RWE-Bayernwerk GmbH" (prospectus issued by KRB); R. Blank, "Der Entwicklungsprozess der Atomwirtschaft," doctoral dissertation, University of Cologne, 1966, pp. 248–259; Schedules of construction and operation startup by correspondence in March and April 1976 with Kernkraftwerk RWE-Bayernwerk GmbH, Kernkraftwerk Lingen GmbH and Kernkraftwerk Obrigheim GmbH. Data on operation losses from an unpublished document of the Federal Ministry for Research and Technology.

[a] The first figure gives the initial rating, the second figure the rating as increased later.

[b] The shares of AEG and of the banks were later taken over by Vereinigte Elektrizitätswerke Westfalen AG.

[c] First operation at full load.

[d] Includes Federal Government and Euratom.

plant in the world, the rating of the plants was increased rapidly to the range of 1,100 to 1,300 MWe in order to take advantage of economies of scale.[65] As a result of a number of factors including a rise in construction costs, a decline in the growth rate of electricity demand, and public opposition to nuclear power, there has been only one domestic order since 1976. At the end of 1979 altogether eight commercial LWR plants with a total capacity of 8,200 MWe were in operation. Domestic orders existed for further thirteen plants of this type, with an aggregate capacity of 16,000 MWe.[66]

The two main West German suppliers for LWRs, AEG and Siemens, joined forces in their electric-power business in April 1969, when they formed the Kraftwerk Union AG, each parent company holding 50 percent of the stocks.[67] Because of existing license agreements with GE and Westinghouse, the nuclear divisions of the two companies were included in the merger only in 1973.[68] Siemens was in a position to terminate its license agreement with Westinghouse in 1970. Since 1972 Kraftwerk Union has cooperated in the field of PWRs with Combustion Engineering (United States). After the Kahl and Gundremmingen plants AEG replaced its GE license by an agreement, so Kraftwerk Union is now able to offer its own design for BWRs.[69] Interatom became a subsidiary of Kraftwerk Union; Siemens had acquired a majority holding in 1969 and, after having taken over the remaining shares, transferred ownership in 1973 to Kraftwerk Union.[70] In 1976, AEG pulled out of Kraftwerk Union, selling its shares to Siemens, so Kraftwerk Union is now fully owned by Siemens.[71]

In addition to domestic orders, Kraftwerk Union or its parent companies bid successfully on foreign markets. Apart from two plants with heavy-water reactors for Argentine, all foreign orders were for LWR plants, mainly of the pressurized-water type. Up to the end of 1979, export orders of Kraftwerk Union amounted to a total capacity of about 15,000 MWe. Two of these orders were part of a West German–Brazilian collaboration agreement. In addition to the construction of a number of nuclear power plants, this agreement provided for the supply of pilot facilities for uranium enrichment and reprocessing, and hence has intensified worldwide discussions about the adequacy of international safeguards against the proliferation of nuclear weapons.[72] Kraftwerk Union is now, in terms of plant orders, one of the world's largest suppliers of nuclear plants. It ranks fourth on the world market after Westinghouse, GE (both United States), and Framatome (France).[73]

As a second German supplier for LWRs, Babcock-Brown Boveri Reaktor GmbH (BBR) was formed in 1971 by BBC, Mannheim (26 percent), the American company Babcock & Wilcox (69 percent), and Deutsche Babcock & Wilcox (5 percent). This company offers PWRs under a Babcock & Wilcox license. It is assumed in nuclear reactor circles that its formation was encouraged by German utilities in order to prevent a monopoly by Kraftwerk Union. Up to the end of 1979, BBR received orders for two generating units from West German utilities. There have been some changes in the distribution of shares in recent years, and since the beginning of 1980 the shareholders are BBC, Switzerland (51 percent) and BBC, Mannheim (49 percent).[74]

Government support for LWRs could be reduced, with this technology getting into its commercial phase. After the appropriations for the three demonstration plants, a further DM20 million of government money was spent in support of R&D for LWRs up to 1972.[75] For the years 1973 to 1976 government plans provided DM57 million for industrial R&D and DM39 million for R&D in government laboratories.[76] Although the technology of LWRs is no longer dependent on government support, substantial government aid is still given to R&D for some of the fuel-cycle components such as uranium enrichment as well as fuel reprocessing and waste disposal.[77]

Government support was increasingly concentrated on HTGCRs and FBRs. In 1968, the government decided to finance a 25-MWe HTGCR designed by GHH under a General Atomic license. GHH drew out, however, from already-agreed construction contracts, as some of the agreed guarantees appeared too risky. The project was abandoned, and the firm reduced its nuclear power plant activities to the manufacture of plant components.[78] In 1970 the government agreed to give financial support for a 300-MWe high-temperature gas-cooled prototype plant, the nuclear system of which was to be supplied by Brown, Boveri/Krupp Reaktorbau GmbH (BBK). Krupp withdrew from the venture, however, before a contract was signed.[79] In 1973, BBC combined for its work on the high-temperature reactor with Gulf General Atomic (United States) as a minority partner in the joint venture Hochtemperatur-Reaktorbau GmbH (HRB).[80] HRB is constructing the 300-MWe THTR-300 plant with an HTGCR at a site near Schmehausen. Operation start is scheduled for 1983. Construction costs are presently estimated at DM1,680 million, plus DM100 million for R&D. Apart from a small fraction paid by the constructor and by some utilities owning and operating the plant, the costs are financed by the Federal Government and the respective Land Government, or through loans guaranteed by the state.[81]

Work on FBRs had begun in West Germany in 1960, when the Karlsruhe laboratory started a three-year program. An association agreed-on with Euratom in 1963 extended the program up to 1967. Design contracts for two 300-MWe fast breeder prototypes, a sodium-cooled and a steam-cooled one, were given by the government to German firms in 1966. The design on the steam-cooled type was terminated in February 1969 in favor of the sodium-cooled type. In 1972 the decision was taken to construct a mainly government-financed 300-MWe sodium-cooled prototype plant SNR-300 in collaboration with Belgium and the Netherlands. The SNR-300 plant, located near Kalkar on the bank of the Rhine, is scheduled for completion in 1985. The government decisions in this program are the subject of the following chapters.

Relative Success

If its outcome is measured against its original objectives, reactor development in West Germany has been a success. West Germany has caught up with other countries, and an independent domestic nuclear industry has been established that is able to compete on the world market. The basic technical choices in the West German reactor program seem to contrast favorably with those of other medium-sized countries, for instance France and Britain. Within the wide variety of technical options available, the West Germans selected early for commercial use the LWR, which is now the dominant type of reactor in the world.[82] France and Britain, in contrast, spent large funds on construction programs for gas-graphite reactors of domestic design that were commercially deployed only in protected home markets.[83] After a construction program of 2,200 MWe comprising eight gas-graphite reactors, the French abandoned their domestic design in favor of a program of LWRs based on an American license.[84] Britain discarded her Magnox design after installing a capacity of about 4,200 MWe with twenty-six reactors; the AGR succeeding the Magnox design seems to be phased out after construction of 8,700 MWe with fifteen reactors, and decisions have been taken to construct some LWRs under a foreign license.[85] The policy reversal in France and Britain seems to imply a recognition that the gas-graphite programs were bad investments from an economic point of view.[86]

However, if the outcome of West German reactor development is appraised with sober economic sense, success must be qualified, whatever criteria are used.[87] Out of the many firms which in 1955 embarked on nuclear reactor development, only the three big West German electrical companies ended up with commercial sales: Siemens and AEG finally independent of foreign know-how, BBC under a license of Babcock & Wilcox (United States). The other firms which had prior experience only in power-plant components or other types of heavy-plant equipment, but not in the construction of whole power plants, gradually pulled out of the reactor program. This applies to Demag, Krupp, MAN, GHH, and Deutsche Babcock & Wilcox.

Like foreign firms engaged in nuclear plants, those West German firms that did make commercial sales have not made, on balance, a profit so far. AEG has recently pulled out of the nuclear plant business with a net loss of DM1,700 million. Siemens's experience was less dismal, though inflationary cost-overruns in some fixed-price contracts also brought red figures in some years. Kraftwerk Union, the joint subsidiary of AEG and Siemens and since 1977 only of Siemens, has produced profits only in 1976 and 1979. BBR also sustained losses. On the firm level, therefore, the commercial success of West German reactor development, as that in other countries, has yet to be established.[88]

What are the contributions of government policy to the relative success of the West German effort? A major contribution is due less to intentional action of government agencies than to the decentralized organizational framework, which set limits for the actions of the ministry in charge of the nuclear program. The choice of reactor type was made not by the government but by supplier firms and utilities within the scope of the licensing requirements imposed by the safety authorities. In designing its early programs, the government deliberately left technical choices to suppliers firms. Moreover, the key decisions as to the choice of reactor type were taken independently of the government programs and, as far as the early policy of the utilities is concerned, against the government. The PWR fueled by enriched uranium, later to be the most successful type of reactor, did not feature in any of the early government programs.

What appears to be *the* West German reactor program, is in fact a conglomerate of several strategies: that of the utilities taking their choice among internationally available technology, and that of the government emphasizing the development of an independent domestic technology. In the conflict between these two strategies, reactor manufacturers tried to reap the benefits of both: they used government finance for developing indigenous technology, at the same time they cooperated with foreign firms in order to be able to offer foreign designs before their own designs were ready for making bids. Those firms which finally were able to sell reactors independently of foreign licenses, Siemens and AEG, added a third strategy: rather than relying on the government program or on foreign technology, they used these as supplements to their own self-financed efforts. Siemens developed the heavy-water pressure-vessel reactor and the pressurized LWR up to complete designs without government support and independently of government programs. The firm's commitment of its own funds up to the first commercial order in 1967 is estimated by knowledgeable sources at around DM100 million.[89] Although the company drew on a Westinghouse license for some reactor components, extending the long-standing cross-license agreement of the two companies, it was basically an independent development effort. AEG's design activities on the BWR were also largely financed from the company's own funds. Its commitment of own funds up to the first fully commercial order in 1967 is estimated by knowledgeable sources as of the same order of magnitude as that of Siemens, that is, on the order of DM100 million.[90] Also some West German manufacturers of reactor components, such as pumps, pressure vessels, and steam generators, invested a good deal of their own money in development work. Relative to their size, some of these firms invested even more than Siemens or AEG.[91]

The involvement of independent domestic firms commanding a strong technical and commercial capability and willing to invest large amounts of their own funds was therefore an essential condition of the relative success of the West German effort. There are doubts whether government subsidies to manufacturers and utilities were a key factor. Those types of reactors which finally were successful were developed with relatively little government money. Design work was largely financed by the supplier firms from their own funds. The utilities were capable of financing the costs of small prototype plants with their own money. Since the light-water demonstration plants had nearly commercial electricity costs, their construction costs could be largely financed by the utilities through stock capital and loans (see table 2–4). The contribution of government laboratories to LWRs was marginal. Their work focused on the HTGCR (Jülich) or the FBR (Karlsruhe). They became involved in LWR technology only in the early 1970s on a small scale, with an emphasis on reactor safety. Out of a total expenditure of DM954 million for the construction of LWR prototype and demonstration plants (VAK Kahl, KRB Gundremmingen, KWL Lingen, and KWO Obrigheim), only DM112 million (12 percent) were financed by the state.[92]

As table 2–5 shows, out of a total of DM2,752 million of government support to reactor development up to 1972,[93] only a minor share (14.0 percent) was spent on those reactor types for which commercial orders have been placed: the PWR, which at present is the favorite reactor type on the world market, received DM66 million or 2.4 percent; the BWR DM122 million or 4.4 percent; the pressure-vessel heavy-water reactor DM199 million or 7.2 percent. The bulk of government expenditure (62.1 percent) went on those reactor types which may have a chance for commercial application only in the long term: DM1,084 million or 39.4 percent on the FBR and DM624 million or 22.7 percent on the HTGCR. A proportion that is not negligible (22.7 percent) was given for reactor types that by now have to be regarded as technical or commercial failures, including the nuclear superheat reactor (DM159 million or 5.8 percent), the heavy-water gas-cooled reactor (DM155 million or 5.6 percent), gas-graphite reactors (DM17 million or 0.6 percent), naval propulsion reactors (DM288 million or 10.5 percent), and the incore thermionic reactor (DM21 million or 0.8 percent).

The case of West German reactor development appears to be consistent with the view that the scale of R&D cost is not a prohibitive factor in technological innovation, if firms are backed by government-financed basic research and exploratory development as well as by government risk sharing (see chapter 1). The DM100 million Siemens spent until 1967 have taken

Table 2-5
Expenditure on Reactor Development by Government and Utilities until 1972
(DM in millions)

	Boiling-Water Reactor [a]	Pressurized-Water Reactor [a]	Nuclear Superheat Reactor [b]	Pressure-Vessel Heavy-Water Reactor [c]	Heavy-Water Gas-Cooled Reactor [c]	High-Temperature Gas-Cooled Reactor	Fast Breeder Reactors [d]	Gas-Graphite Reactor (Magnox and AGR)	Naval Propulsion Reactor [e]	Incore Thermionic Reactor	Other Types, Components, Studies
Government [f]	122	66	159	199	155[l]	624	1084	17	288	21	17
National [g]	90	66	159	199	155[l]	584	966	17	268	21	15
Euratom [h]	32	–	–	–	–	40	118	–	20	–	2
Utilities [i]	231	100	1	10	2	76	9	2	–	–	–
Loans [j]	322	191	–	–	75	–	–	–	–	–	–
Total	675	357	160	209	232	700	1093	19	288	21	17
Precommercial plants											
Name	VAK	KWO	HDR	MZFR	KKN	AVR	KNK I/II		Nuclear ship Otto Hahn		
Rating	15 MWe	328 MWe	23 MWe	50 MWe	100 MWe	13 MWe	18 MWe		38 MWt		
Date of construction [k]	1958–1962	1965–1969	1965–1971	1961–1966	1966–1974	1959–1969	1966–1978		1962–1968		
	KRB 237 MWe 1962–1967					THTR-300 300 MWe 1971–1983	SNR-300 280 MWe 1972–1985				
	KWL 256 MWe 1964–1968										

Sources: G. Schuster, "Deutsches Reaktorprogramm," in *Taschenbuch für Atomfragen 1968*, edited by W. Cartellieri et al. (Bonn: Festland Verlag, 1968), pp. 147–171; "VAK Versuchsatomkraftwerk Kahl GmbH" (prospectus issued by VAK); "KRB Kernkraftwerk RWE-Bayernwerk GmbH," (prospectus issued by KRB); an unpublished document of the Federal Ministry for Research and Technology; correspondence in November 1979 with Kernforschungs-anlage Jülich and Kernforschungszentrum Karlsruhe, in June 1980 with GKSS.

Table 2-5 continued

Notes: The expenditure listed covers R&D for the reactor proper and fuel fabrication. Excluded are front-end fuel-cycle activities such as uranium prospecting, exploration, mining, and milling. Back-end fuel-cycle activities are included only for the high-temperature gas-cooled reactor and the fast breeder reactor, where they account for a small proportion of total R&D. For the other reactor types, R&D on reprocessing and waste disposal is not included.

a DM10 million were distributed at equal shares on the boiling-water reactor and the pressurized-water reactor, as it was not possible to specify to which type this expenditure related.

b Boiling-water reactor with nuclear steam superheating.

c DM11 million for the heavy-water reactor were distributed at equal shares on the pressure-vessel and the pressure-tube type.

d Includes sodium-cooled thermal reactors.

e Included are all investments and operating costs of the Gesellschaft zur Kernenergieverwertung in Schiffbau und Schiffahrt (GKSS). This laboratory has since 1970 taken on R&D outside naval propulsion, so the figures may overestimate expenditure for naval propulsion. Figures for GKSS investments are budget obligations, not actual expenditure.

f Government expenditure covers all government-financed R&D in utilities, manufacturing industry, and government laboratories, including funds appropriated to government laboratories (Karlsruhe with Versuchsanlagen, Jülich, and Geesthacht) via their yearly budgets. The operation of MZFR, HDR, KNK, and KKN are derived from the yearly estimates of the federal budget. The West German shares in OECD projects (Dragon and Halden) are not included.

g Federal and state.

h Only payments to West German projects.

i Expenditure only for precommercial plants, including the costs of construction and (with the exception of VAK) operation. Other R&D activities are included only as far as they become visible in government contracts.

j All loans were guaranteed by the government.

k The year of the order and the year in which the plant was or is to be turned over to the operator. In the case of KKN and HDR the year of final shutdown is given.

l Includes a government loan of DM10 million at no interest.

about 2 to 3 percent of the firm's R&D budget.[94] It seems possible, therefore, that the DM66 million government subsidy for the PWR and the
DM122 million subsidy for the BWR were not decisive. Siemens and AEG
would have had the capability to do it without this subsidy, though development might have taken longer. Siemens might have constructed the Obrigheim plant somewhat later and AEG might have built only one of the two
demonstration plants. We do not know for certain what the firms would
have done without this subsidy, we can only speculate on what they would
have done. The simple fact that a subsidy was given does not imply that
without it the firms would have withdrawn from developing nuclear power
plants. Perhaps it was good that the West German government did not want
to find out and make a test. But probably a different regime of government
support might have done as well. It may be recalled that some influential
actors on the utility side, for example Heinrich Schöller of RWE, proposed
an even more limited role for government support than the one suggested by
Eads and Nelson (see chapter 1).

The following chapters look in more detail at government decisions in
the support of FBRs. Compared to LWRs, FBRs are a more sophisticated
technology involving a larger R&D effort and longer lead times. Will the
evidence in this case also fall on Eads and Nelson's side, or will it support
the Galbraithian line of arguments?

Notes

1. The main published accounts of nuclear energy in West Germany
are by participants or of semiofficial character: R. Gerwin, *Atomenergie in
Deutschland* (Düsseldorf and Vienna: Econ, 1964); K. Winnacker and
K. Wirtz, *Das unverstandene Wunder: Kernenergie in Deutschland* (Düsseldorf and Vienna: Econ, 1975); E. Bagge, K. Diebner, and K. Jay, *Von der
Uranspaltung bis Calder Hall* (Hamburg: Rowohlt, 1957). Only published
sources are used in two articles by B. Moldenhauer, "Politische und ökono-
mische Entstehungsbedingungen der zivilen Atomindustrie," *Blätter für
deutsche und internationale Politik* 20 (1975):741–754; "Die Atomindustrie
in der BRD," ibid., pp. 1087–1104. Three articles by J. Radkau draw on
documents of government advisory bodies, but are confined to isolated
aspects: "Nationalpolitische Dimensionen der Schwerwasser-Reaktorlinie
in den Anfängen der bundesdeutschen Kernenergie-Entwicklung," *Tech-
nikgeschichte* 45 (1978):229–256; "Die Kalkulation des Unberechenbaren:
Zur Entwicklungs- und Wirkungsweise des industriellen Kernenergie-
Interesses in der BRD," *Blätter für deutsche und internationale Politik* 23

(1978):1440–1466; "Kernenergie-Entwicklung in der Bundesrepublik: ein Lernprozess?" *Geschichte und Gesellschaft* 4 (1978):195–222. Some useful material can be found in the doctoral dissertations by R. Blank, "Der Entwicklungsprozess der Atomwirtschaft," University of Cologne, 1966, and J.P. Pesch, "Staatliche Forschungs- und Entwicklungspolitik im Spannungsfeld zwischen Regierung, Parlament und privaten Experten, untersucht am Beispiel der deutschen Atompolitik," University of Freiburg, 1975.

2. A legal analysis is given in M. Willrich, "West Germany's Pledge Not to Manufacture Nuclear Weapons," *Virginia Journal of International Law* 7 (1966):91–100. For the political context see W. Cornides, "Die Pariser Verträge und der weltpolitische modus vivendi," in *Die Internationale Politik 1955. Eine Einführung in das Geschehen der Gegenwart,* edited by A. Bergstraesser and W. Cornides, Jahrbücher des Forschungsinstituts der Deutschen Gesellschaft für Auswärtige Politik (Munich: Verlag R. Oldenbourg, 1958), pp. 163–179.

3. I. Staff, *Wissenschaftsförderung im Gesamtstaat* (Berlin: Duncker & Humblot, 1971); C. Bode, "Möglichkeiten und Grenzen einer Gesetzgebung des Bundes zur Förderung der wissenschaftlichen Forschung," *Wissenschaftsrecht, Wissenschaftsverwaltung, Wissenschaftsförderung* 5 (1972):222–238.

4. H. Gröner, *Die Ordnung der deutschen Elektrizitätswirtschaft* (Baden-Baden: Nomos, 1975), table 4, p. 58 and table 8, p. 70.

5. D. Irving, *The Virus House* (London; Kimber, 1967).

6. "Demag und NAA gründen INTERATOM," *Atomwirtschaft* 3 (1958):37; L.C. Nehrt, *International Marketing of Nuclear Power Plants* (Bloomington and London: Indiana University Press, 1966), pp. 21–22; "Deutsche Babcock & Wilcox . . . " *Nucleonics Week* (19 May 1966):8.

7. "Deutsche Babcock offeriert Calder Hall-Typ," *Atomwirtschaft* 2 (1957):101.

8. "Gutehoffnungshütte Sterkrade AG (Lizenzvertrag mit der General Atomic)," *Atom und Wasser,* no. 94 (16 May 1960):1–2.

9. "Zusammenarbeit Siemens-Westinghouse," *Atomwirtschaft* 3 (1958):37; personal correspondence with Kraftwerk Union, August 1978.

10. H. Schöller, "Kernenergie für die Elektrizitätswirtschaft—Aber wann?," *Atomwirtschaft* 1 (1956):331–332.

11. "Arbeitsgemeinschaft deutscher Energieversorgungsunternehmen zur Vorbereitung der Errichtung eines Leistungsversuchsreaktors," *Atomwirtschaft* 1 (1956):422; "Neugründung: Studiengesellschaft für Kernkraftwerke," *Atomwirtschaft* 2 (1957):341; "Gesellschaft für die Entwicklung der Atomkraft in Bayern mbH," *Atomwirtschaft* 2 (1957):135; Nehrt, op. cit., note 6, pp. 230–235.

12. "RWE-Kraftwerksprojekt," *Atomwirtschaft* 1 (1956):238;

H. Mandel, "Die Planungen des RWE auf dem Atomsektor," *Atomwirtschaft* 1 (1956):332–334; idem, "RWE-Atomkraftwerk Kahl," *Atomwirtschaft* 2 (1957):253–255; Nehrt, op. cit., note 6, pp. 248–254; "RWE erwägt Annullierung des Siedewasserversuchskraftwerks," *Atomwirtschaft* 2 (1957):342.

13. J. Engelhard, "Abschlussbericht über die Errichtung und den Anfahrbetrieb des AVR-Atomversuchskraftwerkes," Forschungsbericht K 72-23 (Bonn: Bundesministerium für Bildung und Wissenschaft, December 1972), pp. 1–3.

14. L. Brandt, "Zehn Jahre friedliche Atomenergieforschung und -technik in der Bundesrepublik Deutschland," in *ATOM-reaktoren, -energie, -forschung in Deutschland. Situation 1965. Überblick 1966* (Sprendlingen: Heinz P. Conté, 1965), pp. 17–22.

15. In accordance with changes in its responsibilities the ministry in charge of nuclear energy was renamed Ministry for Atomic Energy and Water Economy in October 1957, Ministry for Atomic Nuclear Energy in November 1961, Ministry for Scientific Research in December 1962, Ministry for Education and Science in October 1969. In December 1972 this ministry was split into two parts; responsibility for nuclear power has since been with the Ministry for Research and Technology.

16. For official descriptions of the German Atom Commission see H. Lechmann, "Deutsche Atomkommission," in *Taschenbuch für Atomfragen 1959,* edited by W. Cartellieri et al. (Bonn: Festland Verlag, 1959), pp. 9–16; W. Hesse, "Deutsche Atomkommission," in *Taschenbuch für Atomfragen 1960/61,* edited by W. Cartellieri et al. (Bonn: Festland Verlag, 1960), pp. 9–12.

17. German Atom Commission, minutes of 1st session, 26 January 1956, p. 5 (unpublished).

18. The German Atom Commission was disbanded in 1971 and replaced by other advisory bodies; see "BMBW: Reform des Beratungswesens," *Atomwirtschaft* 16 (1971):498; "DAK löste sich auf," *Atomwirtschaft* 16 (1971):557.

19. F.J. Strauss, "Der Staat in der Atomwirtschaft," *Atomwirtschaft* 1 (1956):2–5; "Atomenergieprogramm," *Atomwirtschaft* 1 (1956):230.

20. See Bundesminister für Forschung und Technologie, *Bundesbericht Forschung V* (Bonn: Bundesministerium für Forschung und Technologie, 1975), pp. 229–242.

21. For published sources see *Atom-Informationen,* no. 23 (28 January 1957); "Deutsche Atomkommission," *Atom-Informationen,* no. 112 (16 May 1957):1–2; "Atomkommission billigt 500-MW-Programm," *Atomwirtschaft* 2 (1957):204.

22. "Siemens-Schuckertwerke AG (Deutsche Gesellschaft für Atomenergie e.V.)," *Atom und Wasser,* no. 221 (November 1958):1–2.

23. J. Brandl, "Die Finanzierung der deutschen Atomkraftwerke," *Atomwirtschaft* 3 (1958):3-5; Anonymous, "Im Dickicht der Finanzierung," *Atomwirtschaft* 3 (1959):2-6.

24. Schöller, op. cit., note 10, p. 331; Mandel, op. cit., note 12, p. 333.

25. German Atom Commission, Working Group II-III/1, minutes of 14th session, 22 July 1958 and of 16th session, 5 December 1958 (unpublished).

26. "Elektrizitätswirtschaft . . . ," *Atom-Informationen,* no. 34 (17 February 1958):1-2.

27. Schöller, op. cit., note 10, p. 331.

28. H. Schöller, "Versuchskraftwerk des RWE," *Industriekurier* (2 August 1956):15. Reprinted with permission.

29. German Atom Commission, Working Group II-III/1, minutes of 14th session, 22 July 1958, of 15th session, 7 November 1958, and of 16th session, 5 December 1958 (unpublished).

30. Mandel, op. cit., note 12.

31. "Kernenergieprogramm," *Atom-Informationen,* no. 251 (12 December 1957).

32. L. Armand, F. Etsel, and F. Giordani, *A Target for Euratom* (Brussels, 1957).

33. For a detailed account of Euratom's reactor policy see H.R. Nau, *National Politics and International Technology* (Baltimore and London: Johns Hopkins University Press, 1974); West German policy toward Euratom is analyzed by C. Deubner, *Die Atompolitik der westdeutschen Industrie und die Gründung von Euratom* (Frankfurt and New York: Campus, 1977). For Euratom's research programs see H. Kramer, *Nuklearpolitik in Westeuropa und die Forschungspolitik der Euratom* (Cologne Carl Heymanns, 1976). The account on the following pages differs in some details from Nau's book and in key aspects from the presentation by I.C. Bupp and J.-C. Derian in chapter 1 of their *Light Water: How the Nuclear Dream Dissolved* (New York: Basic Books, 1978).

34. Brandl, op. cit., note 23.

35. "Arbeitsgemeinschaft Versuchs-Reaktor GmbH (AVR), Düsseldorf," *Atom und Wasser,* no. 24 (4 February 1959):1; "Weitere Mitglieder der AVR," *Atomwirtschaft* 4 (1959):354.

36. Engelhard, op. cit., note 13, pp. 53-55, 247-250.

37. "RWE prüft neue Angebote für Kleinkraftwerk," *Atomwirtschaft* 3 (1958):176; "RWE-Auftrag für Kleinkraftwerk an AEG," *Atomwirtschaft* 3 (1958):246; Nehrt, op. cit., note 6, pp. 263-270.

38. *VAK Versuchsatomkraftwerk Kahl GmbH* (brochure issued by Versuchsatomkraftwerk Kahl GmbH, n.d.).

39. Unpublished document of the Ministry for Research and Technology.

40. H. Schöller, contribution to a discussion, in *Arbeitsgemeinschaft*

für Forschung des Landes Nordrhein-Westfalen, no. 93 (Cologne: Westdeutscher Verlag, 1961), pp. 67–68. Reprinted with permission.

41. "Die weiteren Pläne des RWE," *Atomwirtschaft* 2 (1957):380; "Ausschreibungen für erste Atomkraftwerke laufen," *Atomwirtschaft* 3 (1958):127; "Rheinisch-Westfälische Elektrizitätswerk AG, Essen (Reaktorbau)," *Atom und Strom,* no. 10 (15 January 1959):1–2.

42. Quoted in "Rhein.-Westf. Elektrizitätswerk AG (Versuchsatomkraftwerk Kahl GmbH)," *Atom und Wasser,* no. 19 (28 January 1959):1–2.

43. "Fünf Vorschläge im Euratom/USA-Reaktorbauprogramm," "AKS meldet Atomkraftwerksprojekt bei Euratom an," "Auch Bewag im Euratom/USA-Programm," all in *Atomwirtschaft* 4 (1959):496; Nehrt, op. cit., note 6, pp. 223–224, 230–235; "AKS-Projekt zurückgezogen," *Atomwirtschaft* 5 (1960):78; "BMAt (Arbeitsgemeinschaft Baden-Württemberg . . .)," *Atom und Wasser,* no. 24 (4 February 1960):1–2; "Rücktrittsbeschluss bei der AKS," *Atom und Strom* (20 February 1960):22.

44. H. Tebbert, W. Strasser, and H. Plank, "Kernkraftwerk Niederaichbach (KKN) mit gasgekühltem D_2O-Druckröhrenreaktor. 1: Von der Projektierung bis zum Baubeginn," *Atomwirtschaft* 11 (1966):493–497; *Bundeshaushaltsplan 1972,* Comments on Kap. 3105 Tit. 893 10; "KKN: Stillegung," *Atomwirtschaft* 21 (1976):131.

45. A. Ziegler, "The Siemens Multi-Purpose Reactor Design," *Nuclear Power* (March 1961):71–74; Gerwin, op. cit., note 1, pp. 13, 34, 52–54; Winnacker and Wirtz, op. cit., note 1, p. 142. "MZFR erzeugt erstmals Strom," *Atomwirtschaft* 11 (1966):145.

46. *Bundeshaushaltsplan 1968,* Comments on Kap. 3103 Tit. 960c.

47. Interview with a participant, September 1976.

48. H. Kühne, "Betrieb, Bau und Entwicklung von Reaktoren in der Bundesrepublik Deutschland," in *Taschenbuch für Atomfragen 1964,* edited by W. Cartellieri et al. (Bonn: Festland Verlag, 1964), pp. 88–100.

49. "SKW-Babcock-Projekt jetzt für AGR-Typ," *Atomwirtschaft* 5 (1960):36.

50. J. Pretsch, "10 Years of Nuclear Energy Policy in the Federal Republic of Germany," in *State, Science, and Economy as Partners,* edited by J. Sobotta (Berlin and Vienna: A.F. Koska, 1967), pp. 137–147; see also W. Schnurr, Contribution to a discussion, in *Arbeitsgemeinschaft des Landes Nordrhein-Westfalen,* no. 93 (Cologne and Opladen: Westdeutscher Verlag, 1961), pp. 69–71.

51. German Atom Commission, Working Group II-III/1, minutes of 23d session, 6 April 1960; German Atom Commission, minutes of 12th session, 20 April 1960 (unpublished).

52. R. Schulten et al., "Der Hochtemperaturreaktor von BBC/ Krupp," *Atomwirtschaft* 4 (1959):377–384; Engelhard, op. cit., note 13, pp. 89–93.

53. German Atom Commission, Working Group II-III/1, minutes of

23d session, 6 April 1960 (unpublished); Kühne, op. cit., note 48, table after p. 94.

54. German Atom Commission, Working Group II-III/1, minutes of 38th session, 13 May 1963, and 40th session, 1 October 1963; Expert Commission III, minutes of 15th session, 4 December 1963 (unpublished).

55. Interview with a participant, January 1975.

56. Engelhard, op. cit., note 13, pp. 89–93.

57. "HDR: Brennelementschaden," *Atomwirtschaft* 16 (1971):262; "HDR soll Testanlage werden," *Atomwirtschaft* 17 (1972):284.

58. Interview with a participant, January 1975.

59. H. Andrae and W. Marth, "Das Projekt KNK II," *Atomwirtschaft* 18 (1973):93–96.

60. Nau, op. cit., note 33, pp. 144–145; Nehrt, op. cit., note 6, pp. 123–124.

61. H. Mandel, "Das Atomkraftwerksprojekt der Kernkraftwerk RWE-Bayernwerk GmbH," *Atomwirtschaft* 7 (1962):533–535; "Rheinisch-Westfälisches Elektrizitätswerk AG/Bayernwerk AG (Kernkraftwerke)," *Atom und Wasser,* no. 201 (17 October 1961):1–2; K. Peuster, "Kernkraftwerk Gundremmingen," in *ATOM-reaktoren, -energie, -forschung in Deutschland* (Sprendlingen: Heinz P. Conté, 1965), pp. 102–105.

62. Interview with a participant, September 1976.

63. German Atom Commission, Expert Commission III, minutes of 14th session, 19 April 1962; Working Group II-III/1, minutes of 32d session, 2 July 1962; German Atom Commission, 14th session, 11 July 1962 (unpublished). See also: "Bau eines grösseren Kernkraftswerks—Empfehlung der Fachkommision 'Technisch-wirtschaftliche Fragen' der Deutschen Atomkommission," *Atom und Strom* (May 1962):40–41; "Deutsche Atomkommission (14. Sitzung)," *Atom-Informationen,* no. 132 (12 July 1962).

64. German Atom Commission, Working Group II-III/1, minutes of 36th session, 11 February 1963 (unpublished).

65. H. Mandel, "Der Bauentschluss für das Kernkraftwerk Biblis," *Atomwirtschaft* 14 (1969):453–455; idem, "Strukturen der nuklearen Stromerzeugung in den 70er und 80er Jahren," *Atomwirtschaft* 18 (1973): 18–24.

66. "Neue Kernkraftwerke in der Bundesrepublik Deutschland 1980," *Atomwirtschaft* 25 (1980):199–213.

67. For the background of this merger see A.J. Surrey and J.H. Chesshire, *World Market for Electric Power Equipment* (University of Sussex, Science Policy Research Unit, 1972), pp. 136–139.

68. Kraftwerk Union, *Geschäftsbericht 1973,* p. 11.

69. "Kernenergie-Lizenzvertrag Siemens/Westinghouse beendet," *Atomwirtschaft* 15 (1970):311; personal correspondence with Kraftwerk Union, August 1978.

70. "Neue Beteiligungsverhältnisse bei INTERATOM," *Atomwirtschaft* 14 (1969):278-279; "INTERATOM jetzt 100% Siemens," *Atomwirtschaft* 17 (1972):547; "KWU-Geschäftsbericht 1972," *Atomwirtschaft* 18 (1973):380-381.

71. "KWU Becomes 100% Siemens," *Nuclear Engineering International* (December 1976):6; "Siemens Takes Full Control of Kraftwerk Union," *Nuclear News* (January 1977):82.

72. An English version of the agreement is printed in L.M. Muntzing, ed., *International Instruments for Nuclear Technology Transfer* (La Grange Park, Ill.: American Nuclear Society, 1978), pp. 411-418. For American voices see N. Gall, "Atoms for Brazil, Dangers for All," *Foreign Policy,* no. 23 (1976):155-201; W.W. Lowrance, "Nuclear Futures for Sale: To Brazil from West Germany," *International Security* 1 (1976):147-166; E. Wonder, "Nuclear Commerce and Nuclear Proliferation: Germany and Brazil, 1975," *Orbis* 21 (1977):277-306. For West German views see K. Kaiser, "The Great Nuclear Debate: German-American Disagreements," *Foreign Policy,* no. 30 (1978):83-110; L. Wilker, "Das Brasiliengeschäft-ein'diplomatischer Betriebsunfall'?, in *Verwaltete Aussenpolitik: Sicherheits und entspannungspolitische Entscheidungsprozesse in Bonn,* edited by H. Haftendorn et al. (Cologne: Verlag Wissenschaft und Politik, 1978), pp. 191-208; H. Haftendorn, *The Nuclear Triangle: Washington, Bonn and Brasilia: National Nuclear Policies and International Proliferation,* Occasional paper 2 (Georgetown University, School of Foreign Service, June 1978).

73. "atw-Schnellstatistik: Kernkraftwerk 1979—Weltübersicht," *Atomwirtschaft* 25 (1980):157-167.

74. "Babcock-BBC-Reaktorgesellschaft gegründet," *Atomwirtschaft* 17 (1972):11; "Deutsche Babcock verkaufte BBR-Anteile," *Atomwirtschaft* 18 (1973):324; "BBR-Mehrheit an BBC," *Atomwirtschaft* 23 (1978):60; "The U.S. Babcock & Wilcox Company," *Nuclear News* (August 1979):136; Nau, op. cit., note 33, p. 151.

75. Unpublished document of the Federal Ministry for Research and Technology.

76. Federal Ministry for Research and Technology, *Fourth Nuclear Program 1973 to 1976 of the Federal Republic of Germany* (Bonn: Federal Ministry for Research and Technology, 1974), p. 33.

77. Bundesminister für Forschung und Technologie, *Programm Energieforschung und Energietechnologien 1977-1980* (Bonn: Bundesminister für Forschung und Technologie, 1977), pp. 93-135.

78. "Kernkraftwerk Wiesmoor," *Atom und Strom*, no. 2/3 (February/March 1963):25; "Die GHH und Geesthacht II," *Deutscher Forschungsbericht, Sonderbericht Kernenergie* (13 May 1970):92-94.

79. "THTR-Prototyp-Kernkraftwerk wird gebaut," *Pressedienst Bun-*

desministerium für Bildung und Wissenschaft, no. 15/70 (22 July 1970):127; "THTR: Krupp scheidet aus," *Atomwirtschaft* 16 (1971):381.

80. "BBC-Gulf-Vertrag abgeschlossen," *Atomwirtschaft* 18 (1973):64.

81. "Uentrop THTR-300," *Atomwirtschaft* 25 (1980):206–207; *Entwurf Bundeshaushaltsplan für das Jahr 1980, Einzelplan 30,* Comment on Kap. 3005 Tit. 89310.

82. In 1979, 318.9 electrical gigawatts (GWe) of nuclear-power plant were in operation or under construction in the world; of this 267.3 GWe (84 percent) used LWRs (PWRs and BWRs), 13.4 GWe (4 percent) gas-graphite reactors (excluding the HTGCR); see International Atomic Energy Agency, *Power Reactors in Member States* (Vienna: IAEA, 1979).

83. Outside France and Britain only three gas-graphite power plants with a total capacity of 789 MWe were constructed. One 150-MWe plant in Italy and one 159-MWe plant in Japan ordered in 1958 and 1960 from British firms have to be placed in the precommercial phase of nuclear power. A 480-MWe plant in Spain constructed by French firms in 1967 to 1972 is a joint Spanish-French venture. See IAEA, op. cit., note 82; "N.P.P.C. Italian Contract Signed," *Nuclear Engineering* 3 (1958):417; "Japan Signs Contract with G.E.C.," *Nuclear Engineering* 5 (1960):48; "Verträge für Vandellos unterzeichnet," *Atomwirtschaft* 12 (1967):333.

84. Data on nuclear capacity and number of reactors from IAEA, op. cit., note 82. For a brief historical account of government policy see P. Papon, *Le Pouvoir et la Science en France* (Paris: Editions du Centurion, 1978), pp. 103–126 and the related parts in Nau, op. cit., note 33.

85. Data on nuclear capacity and number of reactors from IAEA, op. cit., note 82. For critical accounts of government policy see D. Burn, *The Political Economy of Nuclear Energy,* Institute of Economic Affairs Research Monograph 9 (London: Institute of Economic Affairs, 1967); idem, *Nuclear Power and the Energy Crisis: Politics and the Atomic Industry* (London: Macmillan Press for the Trade Policy Research Centre, 1978); H. J. Rush, G. S. MacKerron, and A.J. Surrey, "The Advanced Gas-Cooled Reactor: A Case Study in Reactor Choice," *Energy Policy* 5 (1977):95–105.

86. P.D. Henderson, "Two British Errors: Their Probable Size and Some Possible Lessons," *Oxford Economic Papers* 29 (1977):159–205.

87. Usually, a development project is said to be a commercial success when it results in sales sufficient to recover the costs and yield a profit. In a more rigorous definition, a new product or process is a commercial success if it yields a rate of return that is higher than the return of alternative investment opportunities. See C. Freeman, *The Economics of Industrial Innovation* (Harmondsworth: Penguin, 1974), pp. 174–175; E. Mansfield et al., *Research and Innovation in the Modern Corporation* (London: Macmillan, 1972), p. 52.

88. AEG-Telefunken, *Geschäftsbericht* (1978): p. 13; "Siemens-Geschäftsbericht 1970/71," *Atomwirtschaft* 17 (1972):239; Kraftwerk Union, *Geschaftsbericht,* 1969–1976); Siemens, *Geschäftsbericht,* (1977 and 1978); "B&W Seeks to Cut Equity in German Manufacturer,"*Nuclear News* (March 1978):70.

89. Interview with a participant, September 1976.

90. Interview with a participant, October 1977.

91. Interview with a participant, November 1977.

92. The construction costs for Kahl, Gundremmingen, Lingen, and Obrigheim were DM49 million, DM304 million, DM270 million, and DM331 million, respectively. Government subsidies were DM0.15 million, DM32 million, DM40 million, and DM40 million, respectively. In this calculation some subsidies, which were formally earmarked for R&D but actually were for construction, are included in the construction costs.

93. Total government expenditure for nuclear R&D in the period 1956 to 1972 amounted to DM11,407 million. This breaks down in DM4,776 million for basic research, DM4,680 million for nuclear technology development, DM350 million for safety and radiation protection, and DM1,601 million for international organizations and other items (Federal Ministry for Research and Technology, op. cit., note 76, p. 20). The figure for nuclear technology development is higher than the total expenditure listed in table 2-5, as this table does not include R&D for uranium prospecting and mining, enrichment, reprocessing of spent fuel, and waste disposal.

94. In 1967, the R&D to sales ratio in West German electrical industry was 6.3 percent. This is the average of all firms reporting R&D activities to a national survey; see H. Echterhoff-Severitt, "Wissenschaftsaufwendungen in der Bundesrepublik Deutschland. Folge 3: Aufwendungen der Wirtschaft für Forschung und Entwicklung im Jahre 1967," *Wirtschaft und Wissenschaft* 17, no. 3 (May/June 1969):19–22, table 5. Siemens, as one of the largest companies, probably had a higher research intensity. On the other hand it is to be assumed that the research intensity was less in previous years. The research intensity of Siemens over the period 1956 to 1967 may, therefore, be reasonably estimated at 5 to 7 percent. Sales of Siemens in the business years 1955–1956 to 1966–1967 were DM62,497 million (see Siemens, *Geschäftsbericht,* 1955–1956 to 1966–1967).

3 Establishing a Fast Breeder Program

The possibility of "breeding" new nuclear fuel while generating electricity was discovered in the American effort to develop the nuclear bomb.[1] After the war, when many government laboratories turned their efforts to civilian applications, the fast breeder reactor (FBR) became an essential part of the beginning civilian nuclear power program in the United States. At Los Alamos Scientific Laboratory a small fast research reactor of 25 thermal kilowatts (kWt), Clementine, started operation in November 1946. The Experimental Breeder Reactor I, designed by Argonne National Laboratory, produced in December 1951 the first electricity derived from the energy of the atomic nucleus. But this plant had only a minute output of 200 electrical kilowatts (kWe). Two larger plants followed in the early 1960s: the Experimental Breeder Reactor II with 20 electrical megawatts (MWe) went critical in 1961, the 66-MWe Enrico Fermi Fast Breeder Reactor achieved criticality in 1963.

Other countries soon followed the U.S. lead (see appendix C). The Soviet Union began work on FBRs around 1948. The BR-2 reactor started operation in 1956 with an output of 100 kWt and in 1959 was brought up to 5,000 kWt. Great Britain started a program around 1951. Its initial focus was construction of the 15 MWe Dounreay Fast Reactor, which achieved full power in 1963.

By the second half of the 1950s, when Germany entered the club of nations developing civilian nuclear power, FBRs had already given up their initial lead to thermal converter reactors, such as light-water reactors, (LWRs), heavy-water reactors, and gas-graphite reactors. These avoided the technical problems associated with the use of fast neutrons and liquid-metal cooling and thus were a less complicated technology. Despite their rather poor utilization of nuclear fuels, LWRs, heavy-water reactors, or gas-graphite reactors promised for the near term to generate electricity more reliably and at less cost than FBRs. Although nuclear power had not yet made the step to commercial use, it was apparent that this step if it was to be made would be made by thermal reactors rather than by FBRs.

While industrial firms in West Germany (as in the United States) therefore concentrated their efforts on various types of thermal reactors, the FBR was chosen by the Karlsruhe laboratory to become the focus of its activities. What were the motivations and economic considerations guiding this choice? As the Karlruhe laboratory was initially a joint government-industry venture, an answer to this question requires a broader look at the relative influences of government and industrial actors in this organization.

Karlsruhe Nuclear Research Laboratory

The initiative for founding the Karlsruhe laboratory came from the Max Planck Institute for Physics at Göttingen, where after World War II a number of German nuclear physicists had assembled under Werner Heisenberg's direction.[2] A team led by Karl Wirtz, one of Heisenberg's coworkers in his wartime work, conducted design studies on a small research reactor fueled by natural uranium. Wirtz obtained support for his team from some industrial firms, which for this purpose formed in November 1954 the Physikalische Studiengesellschaft.

The scientists proposed to construct their research reactor FR2 without foreign technical assistance so domestic manufacturers could gain initial experience in the production of reactor components and materials. Several scientific institutes were to be established which would utilize the reactor for studies in reactor engineering, radiochemistry, and radiobiology.[3]

The Karlsruhe laboratory was founded on 19 July 1956 as a private nonprofit company under the name Kernreaktor Bau- und Betriebsgesellschaft mbH (Nuclear Reactor Construction and Operation Company), with a capital stock of DM30 million.[4] Fifty percent of the stock was held by private industry through the Kernreaktor-Finanzierungs-GmbH (Nuclear Reactor Financing Company), which had an initial membership of sixty-five, and later of ninetytwo firms.[5] The other 50 percent of the stock was held by the public sector: 30 percent by the Federal Government and 20 percent by the Land Baden-Württemberg, on whose territory Karlsruhe is located.

The capital stock was to cover the investment costs of the reactor and of the attached institutes. The total investment was estimated at DM40 million, and the contracts obliged stockholders to increase later their shares accordingly. Operating costs of the laboratory were to be financed by the two governments: the Federal Government and the Land Government, sharing at a ratio of 60 to 40 percent. While payments for operation were to begin with the operation startup of the reactor, any activities not directly related to construction of the reactor were to be financed by the governments even before that date.

Major decisions in the Karlsruhe laboratory were the province of a supervisory board consisting of twelve members.[6] Five represented private industry, three the Federal Government, and two the Land Baden-Württemberg. Two were to be appointed jointly by the stockholders. The two representatives of the Land Government of Baden-Württemberg were usually the Minister of the Economy and the Minister of Finance. The Federal Government was represented by senior civil servants: two from the

Ministry for Atomic Questions and one from the Ministry of Finance.[7] The representatives of private industry were top executives of major companies such as Hoechst (chemicals), Gutehoffnungshütte (heavy-plant equipment), Badenwerk (electric utility), Siemens (electrical industry), and Mannesmann (metal products). The jointly appointed members were a university physicist and a trade union representative.

The reactor group of Wirtz with thirty-six individuals moved to Karlsruhe in October 1956. Soon afterward it received a formal framework in the Institute of Neutron Physics and Reactor Technology, with Wirtz appointed director. In addition to this institute, the Karlsruhe laboratory comprised by December 1957 an Institute for Radiochemistry and an Institute for Radiological Protection and Radiobiology. There was also a technical division, which was in charge of the construction of the FR2 reactor. At that time, 252 individuals were employed by the laboratory, not including forty-six individuals seconded from industrial firms.[8]

Originally the Karlsruhe laboratory was intended to be a "reactor station" like several others founded at that time in Berlin, Geesthacht, or Jülich. The statutes of the Kernreaktor Bau- and Betriebsgesellschaft defined its objective as: "the production, collection and evaluation of scientific and technical knowledge gained in the construction and operation of a reactor station, for the peaceful development and utilization of nuclear energy in the public interest."[9]

Against this rather modest conception, the Atom Ministry and the Land Government of Baden-Württemberg urged a considerable expansion. In 1958, the Atom Ministry allocated the first funds to the Karlsruhe reactor station for work unrelated to construction and operation of the FR2 reactor.[10] The Minister for the Economy of the Land Baden-Württemberg spoke at that time of "the conception of Karlsruhe as the German nuclear research center." He stated that the DM40 million originally planned for Karlsruhe would not be sufficient to sustain that conception: "Everything is being tried, however, so that Karlsruhe does not remain an incomplete creation."[11] At the turn of the year 1958–1959, the Atom Ministry and the government of Baden-Württemberg agreed on an investment program for an additional DM70 million.[12] This expansion was justified by the argument that with the few institutes as planned by the Kernreaktor Bau- und Betriebsgesellschaft the research reactor would not be fully utilized. The finance for the expansion was provided by the Federal Government (75 percent) and the Land Government (25 percent) through the Gesellschaft für Kernforschung, founded in June 1959. On the Land's contribution, a ceiling was placed at DM20 million for investments and at DM7.5 million annually for current expenditures. In the latter figure the Land's contribution to the operating costs of the Kernreaktor Bau- und Betriebsgesellschaft was to be included.

The costs of the FR2 reactor rose considerably above the original estimates, partly due to a decision to increase the reactor's output from 6 to 12 MWt. Industry was prepared to contribute in 1959 to an increase in the capital of the Kernreaktor Bau- und Betriebsgesellschaft up to the DM60 million necessary to cover the cost-overruns. It made clear, however, that it was not going to give money for further investments in the Karlsruhe center planned by the Federal Government and the government of Baden-Württemberg. The Federal Government, in addition to its contribution to the increase in stock, covered part of the cost-overruns by a grant for purchase of the fuel and the moderator.[13]

When cost-overruns increased further in 1960 and 1961, an additional financial contribution from private industry was not on the agenda. To leave the structure of the Kernreaktor Bau- und Betriebsgesellschaft intact, the state gave the required funds indirectly through the Gesellschaft für Kernforschung. This organization took over in 1962 the scientific institutes of the Kernreaktor Bau- und Betriebsgesellschaft against refund of the construction costs of the institute buildings. These funds were used to complete the FR2 reactor.

Finally, industry gave its shares on the Kernreaktor Bau- und Betriebsgesellschaft to the Gesellschaft für Kernforschung, that is, to the government. The two companies were merged in January 1964 under the name Gesellschaft für Kernforschung.[14]

Although founded as a joint venture with private industry, the Karlsruhe laboratory must be classed as a government laboratory. In the Kernreaktor Bau- und Betriebsgesellschaft the government, Federal and Land, already had a predominant role, as it assumed the obligation to finance the operating costs of the reactor and all activities not directly related to construction of the reactor. The change in the objective of the Karlsruhe facility from a reactor station to the West German nuclear research center was pushed primarily by the Federal Ministry for Atomic Questions and the Land Government of Baden-Württemberg. As far as the Land Government is concerned, regional interests may suffice to explain this commitment. For the Federal Atom Ministry the expansion of Karlsruhe was instrumental in assuring itself a permanent role at a time when the Eltville Program (see chapter 2) showed slow progress, when its function of promoting the development of nuclear technology was challenged by competition from the European Atomic Energy Community, and when in the support of nuclear research it had to compete with the Federal Länder.

But was industry's financial contribution to the construction costs of the FR2 not an acknowledgment of the benefits it expected from this reactor? This might have been so only for a few firms. The Kernreaktor-Finanzierungs-GmbH included not only firms with a possible interest in nuclear reactors or nuclear materials, but also a good number of firms which definitely had no direct interest in nuclear technology, such as automobile manufacturers, coal mines, and oil companies. Their contribution must be seen as a sort of donation. Furthermore, the financial contribution of a firm did not establish any rights on the utilization of the reactor. Firms were to pay for use of the reactor or other services of the reactor station, whether or not they were a member of the Kernreaktor-Finanzierungs-GmbH.[15]

Preliminary Studies

After the main work on the reactor physics of the FR2 was completed, the scientists in the Institute for Neutron Physics and Reactor Technology sought an orientation for their future work. Wirtz was aware that the development of thermal reactors was already an affair of industry to which the laboratory could not make a significant contribution. Little nuclear physics research remained to be done on this technology. FBRs, however, posed a number of interesting problems where the physicists could make an essential contribution.[16] A seminar during the winter semester 1957–1958 marks the beginning of FBR studies at the Karlsruhe laboratory. In 1958–1959 first calculations on the interrelations between breeding, the critical mass, and the composition of the reactor core were performed on a Zuse Z-22 computer.[17]

The intentions for a fast breeder project in Karlsruhe were brought to the attention of the advisory bodies of the Atom Commission as early as 1957. Up to the reorganization of the Atom Commission in 1966, the committee in charge of reactor development, Working Group II-III/1 on Nuclear Reactors (henceforth Working Group; see appendix B) was chaired by Wirtz. Hence it was not difficult for the Karlsruhe laboratory to present its concerns. At an intermediate phase of the preparation of the Eltville Program, the Karlsruhe project had been considered for inclusion in the 500 Megawatt Program. This would have implied the construction of a 100-MWe FBR plant up to 1965. However, the final program as approved by the Atom Commission in December 1957 mentioned FBRs only as an advanced technology to be developed after the 500 Megawatt Program, though preparations were to be initiated in parallel to it.[18] Since breeder reactors were regarded as a long-term affair, some members of the Working Group proposed to shift their development to a European level.[19]

The Karlsruhe plans were discussed in more detail by the Working Group in March 1958. When Atom Minister Balke described them to the Working Group, he included among other items the design work for a breeder reactor or a fast reactor. In July the Working Group finally agreed that Karlsruhe might work on advanced reactor concepts.[20]

In 1959 Wolf Häfele, head of the theoretical division of Wirtz's institute, was dispatched for a year to the Oak Ridge National Laboratory

(ORNL) in the United States to gather information on American work on both FBRs and thermal breeder reactors.[21] Upon his return in early 1960, the approval of the supervisory board was obtained to set up a project group through collaboration between the Institute for Neutron Physics and Reactor Technology and the technical division. Although the Karlsruhe laboratory was at that time still a joint venture of state and private industry, the supervisory board's approval implied a financial commitment only for the Federal Government. Industry's contribution was in any case limited to the construction of the FR2 reactor. As to the Land Baden-Württemberg, the ceiling of its contribution had already been reached for investments and was nearly reached for operating costs. For the Federal Atom Ministry, however, the fast breeder project was only a part of its plans for a rapid expansion of the Karlsruhe laboratory.[22]

The establishment of the project group in April 1960 is usually regarded as the beginning of the West German fast breeder project.[23] Häfele, continuing his post at the Institute for Neutron Physics and Reactor Technology, was appointed project manager. The scientists expected their project not only to define new tasks for the institute but to play a central role in the laboratory, integrating a scientific institution otherwise disaggregated into a number of institutes, each led by a different director.[24] Such a new task was particularly urgent for the nearly 100 scientists and engineers in the technical division, which faced the possibility of layoff after the completion of the FR2 reactor.[25] The idea of having the Karlsruhe laboratory organized around a project, and thus introducing a second organizational level in addition to the institutes, was derived from precedents in national laboratories in the United States with which Häfele became acquainted during his stay at ORNL.[26]

As the supervisory board's approval was given upon the condition that the committees of the German Atom Commission give their consent, the fast breeder project was presented to the Working Group on 30 September 1960.[27] Membership in this committee comprised at that time fourteen individuals, one with the status of a permanent guest. The organizational affiliations of the members were as follows. Five were physicists holding university professorships, two of these five having simultaneously senior posts at government laboratories; five were executives of reactor manufacturers, usually the heads of the reactor divisions; one was an executive from a large utility; and three were executives from chemical or chemical engineering companies.

The Karlsruhe project group presented plans for a three-year program with the objective of a preliminary choice of possible types. This meant that all technical options for FBRs, in particular with regard to different options for fuel and coolant, should be explored. At the end of the three-year program one specific type was to be selected, and as a subsequent step a prototype plant was to be designed and constructed. An economic justification for developing FBRs was provided in terms of efficient utilization of uranium, though this rationale was put in a global perspective of world fuel reserves rather than in a national perspective of independence from imports.

The discussions of the Working Group, however, focused on more practical questions: what the costs would be; whether Karlsruhe had the competence for this project; whether too much of Karlsruhe's capacity would be bound up by the project and too little scope would be left for other activities; and whether international collaboration should be sought for the project. For a decision on these questions the Working Group deemed more information necessary from the Karlsruhe group, particularly on costs. Therefore, the Karlsruhe group was requested to submit a detailed cost estimate at a further meeting.

The cost estimate as specified thereupon by the project group at the Working Group's session on 7 December 1960 amounted to DM25 million for three-years' work, of which DM13 million were for investments and materials and DM12 million for personnel.[28] In the following discussions ministry officials advocated the project. For instance, the question was discussed as to whether the fast breeder project might tie up too much of the laboratory's capacity to the detriment of other research programs. Ministry officials proposed that in such an event, government laboratories at Jülich or Geesthacht might take over parts of the project.

The Working Group finally resolved that the project was a worthwhile pursuit and that Karlsruhe had the competence to tackle it. It approved that 70 percent of the capacity of the Institute for Neutron Physics and Reactor Technology as well as of the technical division be devoted to the project. The cost estimate was accepted with the qualification that DM8 million out of the total of DM25 million should be covered from current funds of the laboratory. The project group was charged to report twice a year to the Atom Ministry and the Working Group. With regard to international cooperation, there was agreement that the project should start as a West German effort and collaboration with Euratom or other organization should be considered only later, after some progress had been made. After all, the Working Group explicitly recognized that the three-years' effort could result in the termination of the project.

This somewhat sour reception was due to Working Group members with industrial or university background. Speakers of industry had repeatedly advocated that design and engineering work should be given to industrial firms, but simultaneously expressed their apprehensions that this was not very likely to happen because of the structure of the Karlsruhe laboratory.[29] These discussions reveal dissent as to the basic definition of the objectives of the Karlsruhe laboratory between the Karlsruhe scientists and industry. Industry would have liked to see the laboratory confine itself to a more fundamental type of work, whereas the Karlsruhe scientists wanted to reach out into engineering and reactor design. In these discussions the Atom Ministry took the role not of an impartial arbiter but of a partisan of the Karlsruhe laboratory.

Whereas the Working Group in its early years discussed in detail every proposal for government support in the area of reactor technology, even if the amounts involved were less than DM100,000, the Expert Commission III, Technical-Economic Questions of Reactors, reviewed only the major recommendations of its affiliated four Working Groups (see appendix B). As a rule the Expert Commission approved what a Working Group had resolved. If controversial questions came up, the matter was usually referred back to the Working Group. The different character of the Expert Commission's work is also documented by the frequency of its meetings and by its membership. Whereas the Working Group used to meet about four to five times a year, the Expert Commission usually met only once a year. Membership comprised eighteen individuals: five from reactor manufacturers, five from other manufacturing companies, three from universities, two from government laboratories, two from electric utilities, and one trade-unionist. There were a good number of companies that had members in a Working Group as well as in the pertinent Expert Commission; in such cases the individual in the Expert Commission enjoyed higher rank than the one in the Working Group.

The Karlsruhe fast breeder project was presented to the Expert Commission on 4 February 1961, together with the Program for Advanced Reactors (see chapter 2) and a program for ship reactors. The commission gave its unqualified approval. In a special statement the commission expressed its appreciation that the idea for this project had originated in Karlsruhe and declared its confidence that the researchers there had the necessary competence. Furthermore, it confirmed the Working Group's recommendation that international cooperation should be entered into only in later stages of the program.[30]

Boost from Euratom

The threat of possible termination after three years as imposed by the Working Group did not harass the fledgling Karlsruhe project for long. At the time when the Working Group gave its somewhat cool appraisal of the Karlsruhe project, the European Atomic Energy Community (Euratom) had already begun to show a warm interest in FBRs and was about to offer substantial funds.

As mentioned in chapter 2, Euratom was founded on 1 January 1958 by

Belgium, France, the Federal Republic of Germany, Italy, Luxemburg, and the Netherlands. Its establishment was guided by the political aim of European integration. After the failure of earlier attempts, advocates of European unification had directed their hopes on economic integration in specific industrial sectors. Nuclear power promised to be a suitable sector, as it enjoyed a high reputation and as in a new sector efforts at integration were expected to meet fewer barriers than in traditional sectors. Some proponents also thought that through national efforts alone Western European countries would not be able to catch up with the United States and Great Britain, and that the scale of R&D cost in nuclear power would make individual nations inclined to collaborate on a regional level.

As early as 1959 the Euratom commision had proposed to coordinate and accelerate fast breeder development in the framework of a joint European program.[31] Since Euratom had little success in associating itself with the national R&D efforts on first-generation nuclear power plants (LWRs and gas-graphite reactors), it committed its considerable funds to work on technological niches, such as the organic-cooled reactor, and on advanced designs, such as the FBR.

Of the six member states of Euratom only France was engaged at that time in a substantial fast breeder program. Therefore, the commission proposed to locate the joint program at the French fast reactor laboratory Cadarache. However, a joint program was opposed by France, since the French were ahead of the others. The French program, which had started in 1957, was already preparing for the construction of the small experimental reactor Rhapsodie. Only preliminary studies were being done in Belgium and West Germany, while no work was performed in Italy and the Netherlands. In this situation, a joint program did not promise the French any advantage that could have compensated for giving up national control. To get Euratom involved in fast breeder development despite French resistance to a joint program, the commission picked up the idea of a fast zero-power critical assembly and started preparations to construct such a facility under the auspices of Euratom. The initial idea came from Karlsruhe, but such a facility was not being planned at that time in France or West Germany.[32]

The West Germans had a mixed attitude toward Euratom.[33] The initial positive stance toward the foundation of Euratom reflected mainly the foreign-policy objectives of the Foreign Ministry and the chancellor, whereas the Ministry for the Economy and the Atom Ministry shared the critical attitudes of West German industry. Once Euratom was established, the details of West German policy were determined by the Atom Ministry. For this organization Euratom was a competitor in terms of funds and responsibilities. As part of a policy to curb the power of the commission, the West Germans, backed by other member states, urged Euratom in 1960 to establish a Consultative Committee for Nuclear Research, consisting of national experts, to review the commission's administration of community research programs.[34] For particular fields of research this committee was to form special ad hoc groups, one of which, West Germany insisted, was to deal with FBRs.

As a result of such pressures, the commission gave up its intentions for a single European fast breeder project in favor of a more decentralized approach. Now it envisaged a financial participation in the national programs, among which Euratom would fulfill a coordinating function. At the commission's request, first talks with the Karlsruhe project group and officials of the Atom Ministry took place in the winter of 1960–1961.[35]

By then the French no longer resisted an affiliation of their fast breeder program with Euratom, because they faced serious difficulties in obtaining the required plutonium. The plutonium produced in French reactors was needed for military purposes and thus was not available for civilian purposes. Under the Euratom treaty, however, purchase of plutonium abroad was possible only with the commission's collaboration. Also the zero-power cricital assembly as projected by the commission was of some interest to the French program.

For the Karlsruhe project, an affiliation with Euratom was a question of survival as it was threatened with termination after the three-year preliminary program. But there were additional factors that made an affiliation with Euratom attrative.[36] The critical assembly planned by Euratom promised to be very useful to the project. Euratom's financial participation in the national fast breeder programs forestalled the establishment of a European fast breeder program at the Joint Nuclear Research Centre at Ispra. This idea was lingering in the mind of some commissioners. With the higher wages paid by Euratom, its realization would have drawn personnel away from the Karlsruhe laboratory. Like the French, the Karlsruhe project had difficulties in procuring plutonium, as the civilian power program was in its infancy and no reprocessing of spent fuel was planned for the next years. Furthermore, an association with Euratom would strengthen the position of the fast breeder program inside and outside the Karlsruhe laboratory, as Euratom stipulated certain management structures that enhanced the project manager's weight vis-à-vis the laboratory's managing directors and its governing bodies. After all, collaboration with Euratom carried with it a sort of international acknowledgment and increased the international reputation of the Karlsruhe project.

Negotiations with Euratom were effected by officials of the Atom Ministry and the Karlsruhe project management. The Working Group was regularly informed. Two objectives were deemed most important by the Karlsruhe laboratory: keeping domestic control over the project and having the critical facility planned by Euratom located at Karlsruhe. The question of domestic versus European control was discussed at length by the Working Group in July 1961. A number of conditions were laid down as guide-lines for the coming negotiations: the project management should remain with Karlsruhe and it should have certain competencies with regard to the employment of personnel and the allocation of funds. As the Karlsruhe project was at a preliminary stage, one was afraid that codetermination by Euratom could have had a significant influence in its future directions. The Working Group therefore approved acceleration of the project in order to establish a firm program as a basis for negotiations with Euratom.

The Ministry for Atomic Energy gave its political support to the Karls-ruhe group in the negotiations with Euratom, as the project was the main lever in the ministry's effort to secure a fair return from the substantial West German payments to Euratom. This was a matter of concern for the minis-try, since its own budget in 1961 was only three times larger than West German contributions to Euratom.[37] This concern was even stronger with regard to coming years, as the contributions to Euratom were bound to increase. For the second five-year research program (1963 to 1967), the Euratom commission requested from its member states $480 million, the equivalent of DM1,920 million.[38] Although this sum was reduced to $425 million in the final budget,[39] the 30-percent contribution of the West German government meant an annual expenditure of DM80 million to DM90 million, not including Euratom's administration budget. This meant an increase of annual West German contributions by about 50 percent relative to 1961.

To get what was regarded as a fair return from Euratom, the govern-ment had to present West German projects which were on a similar footing as foreign projects competing for European funds. The Karlsruhe project was the only one which could seriously compete with the French fast breeder program. Therefore it,had to be accelerated and expanded. It was in this context that the supervisory board of the Karlsruhe laboratory deter-mined in May 1961 to increase efforts on the fast breeder project by deploy-ing all appropriate resources and staff, and in 1962 made the fast breeder project the central task of the laboratory.[40] To be sure, it was the state which had the say in the supervisory board, since it had to pay the bill.[41] But in this specific case, the bill carried a premium as it was a means to draw Euratom money back to West Germany which otherwise might have been spent on foreign projects.

With notification of the association agreement in May 1963, the Karls-ruhe project received authorization for a research program encompassing an expenditure of DM185 million for the whole period 1960 to 1967.[42] Euratom was to contribute 40 percent of these funds. The research program added notably to the expansion of the Karlsruhe laboratory, including the establishment of new institutes. The number of scientists, engineers, and technicians directly employed in the fast breeder project grew from 150 in

1962 to 400 in 1966. An additional 300 to 400 were less directly involved in 1966.[43] For the project manager this expansion meant promotion to an institute directorship.

Within the West German reactor program the Euratom association with Karlsruhe effected a revolution of priorities. The fast breeder now received more government support than any other type of reactor. Table 3-1 shows that in the period 1956 to 1967, that is, before the first commercial orders for LWRs were placed in West Germany, DM283 million were spent by the Federal Government on the FBR. This was 31 percent of total funds for reactor development. By comparison, the pressurized-water reactor (PWR) received only DM44 million or 5 percent of the total. As industrial efforts focused on thermal reactors, the priority treatment of fast reactors, that is, of a technology with long-term prospects, hardly squared with the official rhetoric of that time that justified government support for reactor development by the lead of other countries with which West Germany had to catch up.[44] However, representatives of industrial firms in the ministry's advisory bodies had no reason to object to this priority for second-generation technology, since they were assured by government officials that other government programs for reactor development would not suffer because of commitments to the fast breeder.[45]

With regard to the integration objectives of Euratom, the outcome of its fast breeder policy was a limited success. The association with the West German program was paralleled by similar associations with the French and Italian programs. The way negotiations went with the French and German programs encouraged Italy to initiate hastily its own fast breeder program in order to get a share of Euratom's cake. Instead of a unified European fast breeder program, there were now three basically independent national programs. Coordination was limited to exchange of information and plans. Although the association agreements made provisions for seconding Euratom personnel to national projects and for exchanging national staff among the associations, these were utilized only in a few cases for Euratom staff, and not at all for national staff.[46]

The West German demand for domestic control of the project was basically fulfilled. Each association was supervised by a committee, comprised of equal numbers of commission and national representatives. However, national representatives had a veto power if voting as a bloc, as majority decisions of the committee became effective only if they included at least one vote from a national representative. A liaison group was set up to coordinate national programs. It had no decision-making power and its only function was the exchange of information and plans.

A compromise was reached about the critical facilities. Euratom supported construction of two facilities, one at Cadarache and one at Karlsruhe. The 300 kg of plutonium, which Euratom intended to secure from the United States, was divided evenly between the French and the West German

Table 3–1
Priorities in Federal Support to Nuclear Reactor Development, 1956 to 1967

Type of Reactor	(million DM)	(percent)
Pressurized-water reactor	44	4.8
Boiling-water reactor	54	5.9
High-temperature gas-cooled reactor	92	10.1
Nuclear superheat reactor	114	12.5
Pressurized heavy-water reactor	133	14.6
Heavy-water gas-cooled reactor	90	9.9
Fast breeder reactor[a]	283	31.1
Naval propulsion reactor[b]	82	9.0
Gas-graphite reactor	16	1.8
Other reactors and components	3	0.3
Total	911	100.0

Sources: Unpublished document of the Federal Ministry for Research and Technology; *Bundeshaushaltsplan 1956 to 1969;* and correspondence with participants.

Note: Expenditure by the Federal Government for R&D performed by government laboratories, manufacturing firms, or utilities. Excluded is R&D financed by Euratom or the Länder Governments.

[a] Includes sodium-cooled thermal reactors.

[b] All federal grants to the Gesellschaft für Kernenergieverwertung in Schiffbau and Schiffahrt are included.

programs. However, it was agreed that the fuel capsules in both facilities have compatible dimensions. Thus the plutonium could easily be leased between the two programs and, at least for short periods, each facility could mock up a full reactor core.[47]

The German association agreement entailed provisions for the participation of Karlsruhe in a joint reactor project with the U.S. Atomic Energy Commission (AEC), an American utility group, and General Electric (GE): the Southwest Experimental Fast Oxide Reactor (SEFOR). Contracts for this project were signed in May 1964, with Karlsruhe contributing $3 million and Euratom $2 million for an estimated total cost of about $25 million.[48] From the persepective of European collaboration it appears somewhat ironic that collaboration of Karlsruhe with American organizations was closer than with the French or the Italian fast breeder programs.

The expansion of the German fast breeder program through its association with Euratom was done without a fresh look at economic need. In favor of the political aim to secure a fair return from West German contributions to Euratom, the ministry ignored the earlier, rather cool appraisal of the project by its advisory bodies.

Euratom support was also decisive in getting the fast breeder programs in Belgium and the Netherlands started and in linking them with the German program.[49] In Belgium, work on FBRs had started in 1956, when a five-man team was seconded to the American Fermi project by SEEN, a joint study association of several Belgian mining and manufacturing firms. In 1957 SEEN was refounded as an engineering company under the name Belgonucléaire. The experience gained in the Fermi project and in a plutonium laboratory, constructed jointly with the national nuclear research center CEN/SCK, made it possible for Belgonucléaire to procure in 1962 design contracts from the French CEA for a plutonium laboratory, a fast-source reactor, and a zero-power critical facility.

This work, which then formed the major part of Belgonucléaire's business, was financed by the French within their Euratom association. After the facilities were completed, however, the French decided to continue their program independently of other countries and cut ties with Belgonucléaire. However, Belgonucléaire's further existence was assured through an association contract of Euratom with the Belgian government, signed in January 1966. This contract also covered some work on fast breeder fuel in the CEN/SCK nuclear research center. It involved a financial commitment of $4 million up to the end of 1967, with a Euratom share of 35 percent.

In the Netherlands, work on FBRs was started directly upon an offer of funds by Euratom to the Organisatie voor Toegepast Naturwetenschappelijk Onderzoek (TNO) and the Reactor Centrum Nederland (RCN; now the energy research center ECN). The association contract signed in November 1965 implied an authorization of about $3 million up to December 1967. As for RCN, the start of a fast breeder project ended a period of indetermination about its objectives that had resulted from the termination of a program for naval propulsion reactors. For the industrial branch of TNO, which lived partly from government finance and partly from research contracts with industrial firms, the association with Euratom provided project money of an unprecedented scale.

It was Euratom's policy to stipulate that its grants be supplemented by national funds. The politics of securing a fair return from their contributions to Euratom, combined with the lobbying of the receiver organizations, sufficiently disposed the national governments to throw national money after European money.

To prevent further disintegration of European fast breeder activities, the Euratom commission stipulated that the Belgians and Dutch join in with either the French or the West German program. As Belgian and Dutch firms had previously been awarded some design and supply contracts for the French program, France was approached first. But the French program, under the influence of General De Gaulle's hostility toward the European Communities, was oriented toward a more independent national line. Belgium and the Netherlands were offered collaboration on terms which were felt to be very unfavorable or even humiliating, clearly indicating French unwillingness to collaborate.

The West Germans, that is, the Karlsruhe scientists, had a much stronger inclination toward international collaboration and offered much better terms. European integration was a motive in this offer as well as economic arguments for sharing development costs and enlarging the market for the fast breeder. The project manager at Karlsruhe devised for both countries a research program supplementing the West German project. This task was difficult as it also entailed coordination between different institutions within the two countries, which was problematic in the Netherlands because of rivalries between the RCN and the TNO. A formal framework was given to this collaboration through association agreements with Euratom. Although still confined to government laboratories or firms living on government R&D contracts, the Belgian and Dutch collaboration with Karlsruhe entailed certain acknowledgment of the laboratory's competence in fast breeder technology, enhanced its influence in the political decision-making process, and added considerably to its public reputation.

The case of the West German fast breeder association supports the view that international organizations are left to support R&D programs with a low importance to national industry.[50] At the same time that a national advisory committee considered terminating work on fast breeders after a preliminary three-year program, Euratom decided to put substantial funds in this reactor type, affording it the highest priority in its R&D program. The Belgian, Dutch, and Italian fast breeder programs point in the same direction. Moreover, these programs show that the politics of fair return, as involved in an international organization like Euratom, may have a deteriorating effect on national priorities, raising projects of low industrial significance to a high priority in the allocation not only of international but also of national funds.

Perspective on Economic Need

When the West German fast breeder program was established, an economic need in the sense of an existing or anticipated problem to which fast breeder development was a response did not exist. There was ample supply of fossil fuels, particularly of Middle East oil. Domestic coal production was stagnating at an output lower than in the mid-1950s. A shortage in the supply of primary energy was not expected in the near future. Nuclear energy was not yet competitive with conventional electricity generation, and forecasts on the time that it would become competitive were still controversial.

In its discussions in February 1963 with representatives of the larger West German utilities (see chapter 1), the Working Group opined that nuclear power would attain competitive status in the early 1970s. Utilities regarded this schedule as premature.[51] In view of a steady expansion of energy demand, nuclear energy was attributed by some people, including utility speakers, an insurance role for future energy supply. But this was to be achieved not by nuclear energy alone, but through an adequate mix with other primary energy sources and, for some time to come, not through the breeder, but through other types of nuclear plants.

Although speakers from utilities stressed that breeder reactors were to play a role in a later phase of nuclear energy and hence their development was desirable,[52] there was no doubt to people in utilities and manufacturing firms that the FBR would not be a competitor for the first generation of commercial nuclear plants. No breeder reactor plant was under the tenders invited by West German utilities in the late 1950s and early 1960s from foreign and domestic manufacturers. None of the West German manufacturers had selected an FBR for government sponsorship under the 500 Megawatt Program of 1957 or the Program for Advanced Reactors of 1960.

In their presentation to the Working Group in 1960, Karlsruhe scientists based their justification for developing FBRs exclusively on questions of global energy resources.[53] With first-generation nuclear reactors, the energy resources contained in high-grade uranium ore are only a seventh or an eighth of fossil-energy resources. With breeder reactors, however, uranium resources yield eight or nine times more energy than fossil-energy resources. Moreover, as breeder reactors extract much more energy from a given amount of urianum than other nuclear plants, they can economically exploit resources of low-grade uranium ores.

The Karlsruhe scientists did not make an independent projection of when the depletion of low-cost uranium resources would create an economic need for breeder reactors. They recognized that the potential of breeder reactors of achieving a high utilization of uranium would be realized in industrial practice only if breeder reactors operate nearly economically in the near future.

The time scale the Karlsruhe scientists proposed for the West German project was not derived from an assessment of uranium supply, but from the schedules of the American and British fast breeder programs. The scientists pointed out that the U.S. AEC and the U.K. Atomic Energy Authority expected fast breeders to play a significant role for the first time after 1970. In addition, they referred to a review article by an ORNL scientist, who assumed that after 1980 about half of all new nuclear plants would be FBRs. Provided that the West German program proceeded in a timely manner, the Karlsruhe scientists deemed it possible that a West German demonstration plant could be put on line as early as 1970.

Neither the economics of the FBR nor the schedules of foreign programs were discussed explicitly by the Working Group. Its recommendation to consider terminating the Karlsruhe project after three years implied, however, that the majority of the committee members were skeptical about the projections of foreign fast breeder proponents and about the potential of the fast breeder to achieve nearly economic operation in the short-term future.

Given a domestic industrial effort to develop nuclear reactors, a modest state-financed program for exploratory work on FBRs may have made economic sense. Nevertheless, it is important to note that the program schedule proposed by the Karlsruhe scientists was not based on a sober anticipation of economic need but on a supposed competition with foreign state-financed programs. In assessing this competition the scientists were content to rely on the most optimistic expectations of their foreign colleagues.[54]

Development Strategy

Early plans at Karlsruhe divided the program in two sequential phases. In the first phase, the main problems of reactor physics and design were to be investigated. This would lead to the point where a choice could be made on the main technical options available for FBRs, in particular with regard to different kinds of coolant and fuel. One specific design was then to be carried on to the second phase, in which a prototype plant of 200–300 MWe was to be designed and constructed.[55] According to the plans as of 1960, the first phase was to take three years. In the framework of the association with Euratom the program plans were revised so as to extend the first phase to 1967, that is, the year when the association was to expire.

This development strategy differed from that of the American, British, and French programs in two respects. First, the Karlsruhe program chose for its first phase the broadest R&D strategy possible within the province of FBRs, regarding steam and helium as serious coolant options. The foreign programs, by contrast, were focused on liquid-metal cooling. Second, the Karlsruhe program went directly to the construction of a large prototype of 200–300 MWe. The fast breeder programs in the United States, the Soviet Union, Great Britain, France, and Japan all followed the usual engineering practice and built small experimental reactors with outputs in the range 20–100 MWt or 10–20 MWe, before going on to prototypes in the range of 200–300 MWe. Although development strategy is usually considered a merely technical matter, the peculiarities of the Karlsruhe strategy show that in reality there is an interplay between technical and nontechnical factors, including historical accident and organizational self-interest.

Choice of Coolant

The broad development strategy selected by the Karlsruhe program for its first phase was made possible by the technical situation of fast breeder development in the late 1950s. The first FBRs in the United States, the Soviet Union, and Great Britain, constructed in the 1940s and 1950s, used metallic fuel (see appendix C).[56] In the second half of the 1950s it became obvious that metallic fuel had some unfavorable effects which impeded the competitive operation of a metal-fueled fast breeder reactor. At temperatures around 660°C, metallic uranium changes its crystal structure so its volume is significantly increased. This sets limits to the operating temperature and hence to the thermal efficiency. Furthermore, because of some damaging effects of irradiation on metallic fuel, the fuel elements have to be exchanged after a relatively low burnup at 10,000 to 15,000 megawattdays per tonne. This implies frequent reloading of fuel, which, in economic terms, means high costs per unit of electricity through frequent plant shutdown, and through frequent reprocessing and refabrication of the fuel. Hence in the late 1950s in all countries developing nuclear reactors the potential of ceramic fuel, preferably oxide, was investigated. Subsequently metallic fuel was abandoned for various kinds of reactors, not only for the fast breeder. Whereas metal-fueled fast breeder reactors require high power densities which permit cooling only by liquid metal, oxide fuel allows lower power densities and thus opens the option for other coolants such as helium, carbon dioxide, or steam.

It was due to the fresh approach by a group just entering this field and uncommitted to specific options that this situation was fully made use of. At an international conference the project manager stated: "we intend to reexamine every decision formerly made."[57] For Karl Wirtz, a closer look at experience abroad with sodium-cooling for fast reactors had raised considerable doubts about this technology. Hence he decided that his group investigate as thoroughly as possible alternative options for fast reactor coolants.[58] It was a result of such reflections that Häfele, when dispatched for a year to the United States, was sent to the ORNL, which was working on thermal breeder reactors, rather than to the Argonne National Laboratory, which was working on liquid-metal-cooled fast breeder reactors. One of Häfele's missions was to learn what objections existed at ORNL against the liquid-metal fast breeder. Early statements of the Karlsruhe group to the Working Group of the German Atom Commission as well as at international conferences, indeed show substantial skepticism with regard to the liquid-metal fast breeder, in particular to its safety features, and a high willingness to explore other coolants such as helium or carbon dioxide.[59]

The steam-cooled fast breeder did not feature in the initial plans of the Karlsruhe group; its development was taken up as the result of an outside initiative that involved some elements of historical accident.[60] A West German delegation visiting nuclear facilities in Great Britain in 1960 happened to call at the R&D department of the Parsons Company in Newcastle and was surprised to learn that the chief scientist directing this department, Ludolf Ritz, was a German. Ritz, an engineer specialist in aerodynamics, had come to Parsons shortly after World War II. There, among other things, he had worked on gas-cooling for nuclear reactors, for which he took out a number of patents, one of which represented a major contribution to the Magnox reactor program. On the initiative of Otto Hahn, who headed the West German delegation, the Atom Ministry offered Ritz various opportunities for work in West Germany, the most appealing of which was an offer to direct a new institute at the Karlsruhe laboratory.

Ritz had completed a study on a steam-cooled fast breeder reactor at Parsons. Apart from some work in France, of which he learned only later, he was the first to devise this reactor concept. Although this study was received by the management of Parsons Company with some interest, it was not considered by the U.K. Atomic Energy Authority for any further action. When he came to Karlsruhe in 1961, Ritz focused the activities of his new Institut für Reaktorbauelemente (Institute of Reactor Components) on circuit components for steam-cooling, but also worked on components for sodium-cooling. Although the idea and the initiative to explore steam-cooling for FBRs were not those of the Karlsruhe project management, Ritz' work was at that time appreciated by the project management as a valuable contribution to the program. It fitted well in the strategy to explore thoroughly all alternatives to sodium-cooling.

Need for an Experimental Reactor

In early announcements of the Karlsruhe project the importance of test irradiations of fuels and materials was heavily stressed. Initially the intention was to perform such tests in the thermal FR2 research reactor at Karlsruhe.[61] This reactor went critical in 1961, one year after the project was formally established, and achieved full output in December 1962. It was said that such tests together with other experimental and theoretical work at Karlsruhe could substitute for a small test reactor.[62]

Test irradiations in the FR2 reactor began in 1965.[63] But earlier it had become clear to the Karlsruhe scientists that irradiations in thermal neutron flux are of limited significance for FBR fuel elements. The Karlsruhe project, therefore, rented some irradiation space in the Belgian BR2 reactor. This was also a thermal reactor, but with a higher fraction of fast neutrons than the FR2 reactor. Irradiations in the BR2 reactor began in 1964, and in 1968 a new agreement was contracted which for a period of five years reserved 50 percent of the capacity of BR2 for the fast breeder project.[64] As the neutron spectrum of the BR2 reactor is not really representative for a fast reactor, an agreement was made in January 1966 for irradiation tests in the American Enrico Fermi Fast Breeder Reactor.[65] Due to the shutdown of this reactor by an accident in October 1966, these plans did not materialize. By the late 1960s the lack of an appropriate irradiation facility with fast neutron flux had proved a serious bottleneck for the West German–Benelux program.[66] It was only with the help of irradiations in British and French reactors that the West German–Benelux program managed to avoid a delay by a number of years.

Later statements by the Karlsruhe project management dismissed a good deal of the development effort on fuel-element technology up to 1966 as unsystematic and of limited applicability. They acknowledged that more attention should have been paid to fuel-element technology and that provision of a domestic test facility would have been a better strategy.[67] The need for extensive irradiation tests under representative reactor conditions was underlined by the discovery in the 1960s of new deleterious effects on cladding material by irradiation under high temperature and fast neutron flux, which were not known in usual thermal reactors. Embrittlement of cladding material in high temperatures was detected in 1963, and swelling of cladding and structural material in fast neutron flux in 1967.[68] Both discoveries, which were made abroad, necessitated a redirection of development work on fuel-element technology.

Why did the initial strategy of the Karlsruhe project fail to provide for a small test reactor? Some interviewees said that such a reactor was initially considered around 1960–1961, but then was ruled out mainly on account of its safety characteristics.[69] Due to its small size such a test reactor would have encountered some difficulties with regard to the Doppler coefficient. Because of lack of inherent safety it was not deemed licensable in West Germany. It seems, however, that at that time not much effort was spent on devising a licensable concept for a small test reactor. In 1963–1964, when the decision on the construction of the sodium-cooled zirconium-moderated KNK reactor (see chapter 2) was to be taken, the Karlsruhe project management proposed that the reactor be built with a fast core instead of a thermal core.[70] This clearly indicates that those who deemed a small test reactor necessary also thought it possible to find ways to ensure its licensability.

Other interviewees pointed out that in 1960–1961 the FR2 test reactor, the centerpiece of the Karlsruhe laboratory, was just about to assume operation. As reasonable tasks for this reactor were difficult to find, one was happy that the fast breeder program provided an opportunity for the reactor to prove its usefulness.[71] This information is corroborated by the emphasis which the early announcements of the Karlsruhe project laid on the role of the FR2 reactor for irradiation tests.[72]

It may be recalled that around 1960 the government justified the expansion of the Karlsruhe laboratory on the ground that the FR2 could not be fully utilized by the few institutes originally planned. A request for a new fast test reactor would have hardly squared with this justification. For the Karlsruhe laboratory, insistence on early construction of a fast test reactor would have implied recognition that the FR2 reactor was of little use. As late as 1966, publications of the project manager still paid lip service to the contrary by saying that the FR2 reactor "had proved to be particularly favorable" for testing fuel-element samples for FBRs.[73]

The KNK reactor could have provided the required test facility if it had been built straightaway with a fast core instead of a thermal one. The zirconium-moderated version of this plant had no commercial perspective. With a thermal zirconium-hydride-moderated core, the value of this reactor for the fast breeder program was limited. It could yield useful experience in constructing, licensing, and operating sodium-cooling systems, but irradiations in the reactor were of very little use for the fast breeder project.

Why then was the KNK reactor not built straightaway with a fast core? The KNK reactor was an industrial (though government-financed) project and building it with a fast core would not have created a conflict with the official justification for the expansion of the Karlsruhe laboratory. As a matter of fact, the management of Interatom, the firm that had designed the KNK reactor, did mention in 1963 to the Karlsruhe scientists the possibility of constructing the KNK reactor immediately with a fast core.[74] However, influential individuals at Karlsruhe rejected emphatically this idea. They saw the possibility of industry overtaking the Karlsruhe laboratory in the development of fast breeders, thus shifting the focus of the fast breeder program from Karlsruhe to Interatom. Although the project manager made some efforts toward encouraging Interatom to equip the KNK reactor with a fast core, the strong position of other senior scientists set clear limits to what he could do in this regard.

On the other hand, a start with a fast core would have delayed construction of the KNK plant by a few years.[75] This was not in the interest of the Interatom management. Although the Working Group had approved the construction of the KNK project in December 1963,[76] the Science Ministry had difficulties in getting the required budget allocations.[77] A change in the design necessitating delay would not have helped in these difficulties.

The politics surrounding the KNK reactor thus show a well-calibrated balance of organizational interest. The Karlsruhe laboratory gave political support for the KNK project with a thermal core as proposed by Interatom. It wished the plant to be constructed at its site. Prodded by Karlsruhe, a regional utility agreed to take the electricity to its grid and to make a financial contribution of DM4 million, an amount sufficient to cover the costs of the turbogenerator. On Karlsruhe's assurance that the reactor was a necessary part of the fast breeder program, the Working Group changed the location of the reactor from Jülich to Karlsruhe.[78] In fact, Karlsruhe's support was decisive in getting government finance for this reactor. Although the Karlsruhe project management was very much interested in starting the plant with a fast core, it did not propose this as a possibility to the Working Group when government support for the KNK plant was discussed. In 1964, the Karlsruhe project management contracted a DM0.1 million study to Interatom for investigation of the possibility of converting the KNK plant into a fast reactor. As a result of this study, some alterations were made in the design providing for later operation with plutonium fuel and fast neutrons.[79]

A proposal for the design and supply of a fast core was made by Interatom not before late 1966, that is, after the firm had taken the contract for construction of the KNK plant with a thermal core.[80] Although government gave funds for a fast core in 1968,[81] the fast-core version of KNK did not start full operation until 1979. Technical problems were the main cause for this delay. But what prevented West Germany in the first instance from constructing a fast test reactor in a timely manner, were not technical factors but organizational self-interest.

Notes

1. For historical accounts of fast breeder programs in various countries see U.S. Atomic Energy Commission, *Environmental Statement: Liquid Metal Fast Breeder Reactor Program,* Report WASH-1535 (Washington, D.C.: U.S. Government Printing Office, 1974), pp. 2.2-1–2.2-15; W. Häfele et al., "Fast Breeder Reactors," *Annual Review of Nuclear Science* 20 (1970):393–434; R. Rockstroh, "Die Entwicklung schneller Reaktoren," *Kernenergie* 15 (1972):73–77, 213–219.

2. For an account by participants see K. Winnacker and K. Wirtz, *Das unverstandene Wunder: Kernenergie in Deutschland* (Düsseldorf and Vienna: Econ, 1975), pp. 62, 67, 116–141.

3. "Physikalische Studiengesellschaft und Projekt Karlsruhe," *Atomwirtschaft* 1 (1956):35.

4. R. Greifeld, "Die Kernreaktor Bau- und Betriebs-Gesellschaft Karlsruhe," *Atomwirtschaft* 1 (1956):293–294.

5. Winnacker and Wirtz, op. cit., note 2, p. 131.

6. "Gründung der Kernreaktor Bau- und Betriebs-GmbH," *Atomwirtschaft* 1 (1956):278; W. Cartellieri et al., eds. *Taschenbuch für Atomfragen 1960/61* (Bonn: Festland Verlag, 1960), p. 367.

7. Initially one of the places from the Ministry for Atomic Questions was filled by the minister himself. However, as federal statutes forbid a minister to sit on the supervisory board of private companies, he had to give his place to one of his civil servants, but participated in the sessions of the

supervisory board as a guest (see Winnacker and Wirtz, op. cit., note 2, pp. 131–132).

8. R. Greifeld et al., "Der Aufbau der Kernreaktor Bau- und Betriebs-Gesellschaft," *Atomwirtschaft* 2 (1957):394–398.

9. Greifeld, op. cit., note 4, p. 293.

10. *Bundeshaushaltsplan 1958,* Kap. 3102 Tit. 640.

11. "Kernreaktor Bau- und Betriebs-GmbH Karlsruhe," *Atom-Informationen,* no. 56 (19 March 1958):1–2.

12. "Deutsche Gesellschaft für Atomforschung mbH in Vorbereitung," *Atomwirtschaft* 4 (1959):35; "Gesellschaft für Kernforschung mbH, Karlsruhe," *Atom und Wasser,* no. 35 (19 February 1959):2; *Bundeshaushaltsplan 1959,* comments on Kap. 3102 Tit. 895 and 896. The investment program was in later years continuously increased. In the 1961 budget it reached DM115 million (see *Bundeshaushaltsplan 1961,* comments on Kap. 3102 Tit. 640, 641, and 896).

13. Winnacker and Wirtz, op. cit., note 2, pp. 135–136; "Gesellschaft für Atomforschung mbH, Karlsruhe," *Atom und Wasser,* no. 35 (19 February 1959):2.

14. *Bundeshaushaltsplan 1962,* comments on Kap. 3102 Tit. 960a; *Bundeshaushaltsplan 1963,* comments on Kap. 3102 Tit. 640a; *Bundeshaushaltsplan 1964,* comments on Kap. 3103 Tit. 640b.

15. This interpretation does not propose that the expansion of the Karlsruhe laboratory was contrary to wishes of industry, but it does refute the interpretation of other authors viewing the early history of the Karlsruhe center as an example of purposive tactics by which industry shoved onto the state the costs of facilities of which it was the main beneficiary; see K. Prüss, *Kernforschungspolitik in der Bundesrepublik Deutschland* (Frankfurt am Main: Suhrkamp, 1974), p. 38.

16. Interview with a participant, December 1975.

17. W. Häfele, "Das Projekt Schneller Brüter," *Atomwirtschaft* 8 (1963):206–209; W. Häfele and P. Engelmann, "Das Projekt Schneller Brüter," in *10 Jahre Kernforschungszentrum Karlsruhe,* edited by Gesellschaft für Kernforschung (Karlsruhe: Gesellschaft für Kernforschung, 1966), pp. 17–27.

18. German Atom Commission, Working Group II-III/1 on Nuclear Reactors (henceforth Working Group), minutes of 9th session, 29 March 1957; German Atom Commission, Expert Commission III, "Technical-economic Questions for Reactors" (henceforth Expert Commission), minutes of 7th session, 2 May 1957; German Atom Commission, minutes of 8th session, 16 May 1957; "Memorandum der Deutschen Atomkommission zu technischen, wirtschaftlichen und finanziellen Fragen des Atomprogramms," 9 December 1957 (unpublished).

19. Atom Commission, appendix 2, minutes of 7th session, 21

February 1957, p. 3; Working Group, minutes of 10th session, 5 July 1957, p. 13 (unpublished).

20. Working Group, minutes of 13th session, 11 March 1958, pp. 7–8, and minutes of 14th session, 22 July 1958, pp. 16–17 (unpublished).

21. Interview with a participant, December 1975.

22. Total investment planned for the Gesellschaft für Kernforschung as of 1960 was DM94.5 million, of which the Land was to contribute DM20 million. The budget for operating costs was DM20 million, of which the Land contributed DM5 million. See *Bundeshaushaltsplan 1960*, Kap. 3102 Tit. 640, 641, and 896.

23. The exact date at which this project group began to work varies a little in the literature. Häfele later gives 1 April 1960; see W. Häfele, "Entwicklungstendenzen bei Schnellen Brutreaktoren," *Atomwirtschaft* 17 (1972):378–384. Engelmann gives the date of May 1960, see P. Engelmann, "Der Schnelle Brüter," *Atomkernenergie* 9 (1964):335–343.

24. Interview with a participant, March 1975.

25. "Finanzierungsprobleme in Karlsruhe," *Atomwirtschaft* 3 (1958): 518–519.

26. K. Wirtz, "Zur Entwicklung gasgekühlter Schneller Brüter," *Atomwirtschaft* 14 (1969):216–218.

27. Working Group, minutes of 25th session, 30 September 1960, p. 9 (unpublished).

28. Working Group, minutes of 26th session, 7 December 1960 (unpublished).

29. Working Group, minutes of 13th session, 11 March 1958, p. 7; minutes of 26th session, 7 December 1960, p. 5 (unpublished).

30. Expert Commission, minutes 13th session, 4 February 1961, p. 7; The Atom Commission itself with a membership at that time of twenty-seven individuals used to meet only once a year to receive the reports of the Expert Commission and to discuss general aspects of state support to nuclear science and technology. It was informed about the Karlsruhe fast breeder project at its 13th session, 21 April 1961.

31. For information on the politics of the Euratom commission and of the French Commissariat à l'Energie Atomique the following presentation relies on H. Nau, *National Politics and International Technology: Nuclear Reactor Development in Western Europe* (Baltimore and London: Johns Hopkins University Press, 1974), pp. 212–235.

32. Interview with a participant, March 1975.

33. See C. Deubner, *Die Atompolitik der westdeutschen Industrie und die Gründung von Euratom* (Frankfurt and New York: Campus Verlag, 1977).

34. Nau, op. cit., note 31, 216–217.

35. Working Group, minutes of 27th session, 27 February 1961, p. 7 (unpublished).

36. Working Group, minutes of 27th session, 27 February 1961, p. 6; minutes of 28th session, 5 July 1961, p. 8; minutes of 30th session, 26 January 1962, p. 12; interview with a participant, March 1975 (unpublished).

37. The federal budget of 1961 allocated nearly DM60 million to Euratom, whereas the ministry's own budget (without contributions to Euratom and other international organizations) amounted to DM191.5 million. (*Bundeshaushaltsplan 1961*, Einzelplan 31 and Kap. A 6006 Tit. 623 and 624).

38. "A $400-Million Budget for Euratom's Second Five Year Plan," *Nucleonics Week* (1 February 1961):5.

39. *Bundeshaushaltsplan 1963*, comments on Kap. 6006 Tit. 624.

40. Working Group, appendix 7, minutes of 29th session, 9 October 1961, p. 1 (unpublished); Häfele and Engelmann, op. cit., note 17, p. 18.

41. Winnacker and Wirtz, op. cit., note 2, pp. 132–133.

42. A. de Stordeur, "The European Community Activities in the Fast Breeder Field," paper presented at the VIII Congresso Nucleare, Rome, Italy, 1963; W. Häfele, "Das Projekt Schneller Brüter," in *Taschenbuch für Atomfragen 1964*, edited by W. Cartellieri et al. (Bonn: Festland Verlag, 1964), pp. 27–31. The expenditure for the program performed under the Euratom association was later increased, and finally the expenditure over the period 1960–1968 amounted to DM282.6 million, of which Euratom financed DM119 million; see G. Schuster, "Deutsches Reaktorprogramm," in *Taschenbuch für Atomfragen 1968*, edited by W. Cartellieri et al. (Bonn: Festland Verlag, 1968), pp. 147–171. Another article by Schuster gives a figure of DM366 million; see *Die reaktortechnische Entwicklung in der Bundesrepublik Deutschland*, Berichte der Studiengesellschaft zur Förderung der Kernenergieverwertung in Schiffbau und Schiffahrt, no. 16 (Munich: Karl Thiemig, 1968), p. 27.

43. W. Häfele and K. Wirtz, "Ansätze zur Entwicklung eines schnellen Brutreaktors," *Atomwirtschaft* 7 (1962):557–559; W. Häfele, "Schnelle Brutreaktoren, ihr Prinzip, ihre Entwicklung und ihre Rolle in einer Kernenergiewirtschaft," in *Arbeitsgemeinschaft für Forschung des Landes Nordrhein-Westfalen, Natur-, Ingenieur- und Gesellschaftswissenschaften*, no. 163 (Cologne and Opladen: Westdeutscher Verlag, 1966), pp. 49–74.

44. For example, "Atomprogramm der Bundesrepublik Deutschland 1963–1967," in *Taschenbuch für Atomfragen 1964*, pp. 163–200.

45. Working Group, minutes of 30th session, 26 January 1962, p. 14 (unpublished).

46. Nau, op. cit., note 31, pp. 218–220.

47. Working Group, minutes of 30th session, 26 January 1962, p. 13 (unpublished). The first exchange of plutonium took place in 1968, when France leased 90 kg from the stock of its critical facility to Karlsruhe; see W. Häfele, "Statusbericht 1970 des Projekts Schneller Brüter," *Atomwirtschaft* 15 (1970):171–175.

48. These costs include construction of the reactor and the associated R&D program. See "The SEFOR Contract," *Nuclear News* (June 1964):29. For more information on the institutional framework of this project see W. Schnurr and J.R. Welsh, "The SEFOR Reactor—Aspects of International Cooperation," Paper P/533, *Third United Nations International Conference on the Peaceful Uses of Atomic Energy, Geneva 1964,* proceedings, pp. 349–355.

49. The following is based on interviews with participants in Belgium and the Netherlands, October and November 1977; see also "Belgium, Holland Assigned Euratom Fast Reactor Work," *Nucleonics* (February 1966): 28; "All EAE Community Members now in Euratom Program," *Nuclear News* (February 1966):10.

50. K. Pavitt, "Technology in Europe's Future," *Research Policy* 1 (1972):210–273.

51. Gedanken für eine Diskussion über die Entwicklung der Atomenergie für die Zwecke der Elektrizitätswirtschaft in der Bundesrepublik Deutschland (paper prepared by members of Working Group for its discussion with representatives of utilities on its 36th session, 11 February 1963); Stellungnahme der Elektrizitäts-Versorgungsunternehmen zur Frage der Förderung der Atomkraftwerksentwicklung der Bundesrepublik (paper summarizing the views of the representatives of nine large utilities put forward in the same discussion).

52. H. Mandel, "Die Planungen des RWE auf dem Atomsektor," *Atomwirtschaft* 1 (1956):323–334. For similar statements by other utility spokesmen see Expert Commission, minutes of 5th session, 26 November 1956, pp. 5–6; Working Group, appendix 4, minutes of 7th session, 19 December 1956, p. 1 (unpublished).

53. Working Group, appendix 3, minutes of 25th session, 30 September 1960 (unpublished).

54. The statements by the Karlsruhe scientists on expectations and program schedules in the United States and Great Britain made an overly optimistic interpretation of the sources they cite; compare "U.S. Civilian Power Reactor Program, Part IV. Plans for Development," *Atomic Industry Reporter* (17 February 1960):1–14; U.K. Atomic Energy Authority, *Sixth Annual Report 1959/60* (London: HMSO, 1960), p. 14; J.A. Lane, "Economics of Nuclear Power," *Annual Review of Nuclear Science* 9 (1959): 473–492.

55. German Atom Commission, Working Group, appendix 3, minutes

of 25th session, 30 September 1960, and appendix 2, minutes of 26th session 7 December 1960 (unpublished); Häfele, op. cit., note 17, p. 206.

56. See references in note 1 and Häfele and Engelmann, op. cit., note 17.

57. W. Häfele, in *Physics of Fast and Intermediate Reactors,* proceedings of a seminar held in Vienna, 3–11 August 1961 (Vienna: International Atomic Energy Agency, 1962), p. 601.

58. Interviews with participants, December 1975, October 1977.

59. Working Group, appendix 3, minutes of 25th session, 30 September 1960 (unpublished); Häfele, in op. cit., note 57, pp. 544–546, 601–602.

60. Interview with a participant, November 1975; Working Group, appendix 7, minutes of 29th session, 9 October 1961, p. 2 (unpublished). For published information see L. Ritz, "Arbeiten über die Dampfkühlung," in *10 Jahre Kernforschungszentrum Karlsruhe,* edited by Gesellschaft für Kernforschung (Karlsruhe: Gesellschaft für Kernforschung, 1966), pp. 86–97.

61. Working Group, appendix 3, minutes of 25th session, 30 September 1960, p. 8; Häfele, in op. cit., note 57, p. 602; idem, op. cit., note 17, p. 208.

62. Häfele, op. cit., note 42, p. 29.

63. H.E. Häfner, "Brennstoffbestrahlung mit Kapselversuchseinsätzen im FR2-Reaktor," *Atomwirtschaft* 17 (1972):314–315.

64. W. Häfele et al., "The Karlsruhe Fast Breeder Project," Paper P/539, *Third United Nations Conference on the Peaceful Uses of Atomic Energy, Geneva 1964,* proceedings, p. 75; idem, op. cit., note 47, p. 173.

65. "EFFBR to Irradiate European Fuel Elements," *Nuclear News* (February 1966):10; "Euratom bestrahlt Brennelemente für Schnelle Brüter in den USA," *Deutscher Forschungsdienst, Sonderbericht Kernenergie* (13 January 1966):10.

66. W. Häfele, quoted in "Der schnelle Brutreaktor mit Natriumkühlung," *Neue Züricher Zeitung* (13 March 1969).

67. G. Karsten, "Die Arbeiten zur Brennelemententwicklung," *Atomwirtschaft* 16 (1971):256–259; Wirtz, op. cit., note 26, p. 217; W. Häfele, "On the Development of Fast Breeders," *Reactor Technology* 13 (1969/70):18–35.

68. Häfele, op. cit., note 67, p. 25.

69. Interviews with participants, November 1974, November 1975.

70. Interviews with participants, March 1975, October 1977.

71. Interviews with participants, November 1975, March 1980.

72. See note 61.

73. Häfele, op. cit., note 43, p. 64; see also Häfele, op. cit., note 47, p. 173.

74. Interview with a participant, March 1980.

75. Interviews with participants, November 1974.

76. Working Group, minutes of 42d session, 13 December 1963 (unpublished).

77. "Budget Cuts May Delay 3 German Nuclear Projects," *Nucleonics Week* (29 July 1965):6; "Mangelnde Staatshilfe gefährdet das deutsche Atomprogramm," *Handelsblatt* (1 October 1965); "Und Deutschlands Schnelle Brüter?" *Sonntagsblatt* (10 October 1965).

78. Working Group, minutes of 38th session, 13 May 1963, pp. 5–6; and minutes of 40th session, 1 October 1963, pp. 5–6; Expert Commission, minutes of 15th session, 4 December 1963, pp. 11–12 and appendix 1, pp. 1–2; Working Group, minutes of 42d session, 13 December 1963 (unpublished).

79. Interview with a participant, March 1975.

80. Working Group, minutes of 54th session, 14 November 1966, p. 9. See also Expert Commission, appendix 1, minutes of 20th session, 12 January 1967, p. 2 (unpublished). The government issued the appropriation statement for KNK on 23 May 1966; J. Pretsch in *Arbeitsgemeinschaft für Forschung des Landes Nordrhein-Westfalen, Natur-, Ingenieur- und Gesellschaftswissenschaften,* no. 163 (Cologne and Opladen: Westdeutscher Verlag, 1966), p. 87.

81. Interview with a participant, February 1975.

4 The Start of Industrial Research and Development

According to West Germany's industrial policy the manufacture of nuclear power plants is the task of private industry. Government support for nuclear research and development (R&D) aimed at bringing a number of West German firms into a position where they could offer and supply nuclear reactor plants on a commercial basis. Although the fast breeder program was initially confined to a government laboratory, the Karlsruhe project management clearly recognized that at some stage of the project the know-how was to be transferred to industrial firms. The question was when this was to be done and how responsibilities would be shared during this transfer.

Proposal for Industrial Design Studies

During the first phase of the project, which had the objective of making an informed choice among the main technical options, leadership of the program was assumed by the Karlsruhe laboratory. The choice of coolant came down soon to the alternative sodium or steam, as the gas-cooled fast breeder ran into technical difficulties. Although a conceptual study showed helium-cooling to be technically feasible in principle, a reactor vessel accommodating safely the very high pressures required was not available at that time. Appropriate cladding material was not at hand for the fuel rods capable of withstanding high temperatures and of achieving a high burnup. Hence design studies on gas-cooled fast breeder reactors were abandoned altogether after 1963, though some experimental work continued.[1]

At the United Nations Geneva conference in May 1964, where the West German project was presented the first time to a wider international audience, the Karlsruhe scientists proposed that a choice be made between sodium and steam because of the associated reactor physics, in particular the Doppler coefficients.[2] This meant that the criteria of choice were within the main area of competence of the Karlsruhe laboratory. However, it was soon recognized that reactor physics alone did not provide a sufficient basis for this choice. Each design had its pros and cons with regard to the power coefficients. Therefore no one design emerged as definitely superior to the others. Discussions at the Geneva conference and ongoing cost studies for a conceptual design of a sodium-cooled fast breeder plant convinced the

Karlsruhe group that the decision among the design options had to be made on the basis of their economic potential, that is, their prospective electricity costs. The crucial factor for this decision would be the capital-cost component rather then the fuel-cycle-cost component, as fuel-cycle costs were not expected to vary greatly among the different fast breeder types.

For estimating the capital costs of the various breeder types the Karlsruhe project management intended to utilize the expertise of industrial firms. Under contract with Karlsruhe, firms were to draft preliminary designs of the different types of fast breeders according to specifications given by the Karlsruhe group. On the basis of these designs they were to make estimates of plant costs. The studies were to be completed in summer 1967, so that the choice among the coolant options could be taken on schedule toward the end of 1967.[3] Four to six firms were to be involved in these studies; that is, practically all firms in the reactor industry. The contracts were thus a move to coopt industry without giving up the direction of the program.

Although the proposal was approved by Working Group II–III/1 on Nuclear Reactors (henceforth Working Group) in November 1964, it was never realized. Events had taken place abroad which in the view of the Karlsruhe project management necessitated a drastic change in the development strategy.

Acceleration

In September 1964 General Electric (GE; United States) announced the company was confident that it would offer a commercial FBR as early as 1974.[4] This was something of a shock for the Karlsruhe scientists, as commercial introduction of the FBR was now to come much earlier than they had assumed.[5] They had scheduled a 1,000-MW demonstration plant to start operation in 1980 and the first commercial order to be placed the same year (see table 4-1). On this schedule, they had believed, the West German program would reach commercialization at the same time as programs abroad. Now it appeared as if GE would rush far ahead and would start by the mid-1970s a sales attack on the international market. For such an attack the West German program would be badly armed if it proceeded as scheduled. The Karlsruhe group therefore felt that the program had to be pushed ahead as fast as possible, lest it fall hopelessly behind the Americans. It seemed appropriate to enter right away phase two of the project, with design and construction of a prototype. As the decision between the sodium-cooled and the steam-cooled fast breeder could not be taken at that stage and as a reasonable decision seemed in any case quite difficult, a virtue was made of necessity. The proposal was made to design and construct a prototype plant of each type.

The question as to what extent industry should become involved in the prototype phase triggered a minor controversy in the Karlsruhe laboratory. Some wanted the direction of the work on the prototypes to remain with the center and the center to assume the responsibility of the architect-engineer. This had been the arrangement for design and construction of the FR2 research reactor. However, the opinion prevailed that the laboratory, while retaining the overall direction of the program, should commission design to industry, and that industrial firms should later serve as main contractors for construction of the prototype plants. For the scientists making this proposal, experience with the FR2 had shown that there were problems when industrial firms acted only in the role of subcontractors.

The decision to enter the prototype phase with two designs, the steam-cooled and the sodium-cooled, was also suggested by other considerations. There existed two reactor projects that had originated outside the fast breeder program and now could be utilized for it. The sodium-cooled zirconium-moderated KNK reactor designed by Interatom could provide experience on sodium-cooling components and later, with the thermal core replaced by a fast one, could serve as a test facility for irradiation of fuel elements for the sodium-cooled fast breeder. The HDR boiling-water reactor with nuclear superheating projected by AEG could yield experience in irradiation under high-temperature steam and later, when equipped with a mixed fast-thermal core, be used for irradiation tests of fuel elements for the steam-cooled fast breeder. Both reactor projects were part of the Program for Advanced Reactors of 1960 (see chapter 2). Consideration of breeding had played no role, when the ministry in 1960 had commissioned the design contracts for these projects. By 1964 the design phase had been completed, and decisions had to be taken whether the government would or would not finance construction of the reactors.

The new proposal by the Karlsruhe project management for an immediate start on phase two of the fast breeder program with design and construction of two prototype plants was presented to the Working Group in March 1965.[6] In its presentation, the Karlsruhe project management stressed the strong international competition as visible in the September 1964 statement by GE. It also reported intentions of GE and Westinghouse to arrive at a fast breeder prototype plant by 1968. With regard to the participation of industrial firms, speakers of the Karlsruhe project deemed it desirable that as many firms as possible should be involved. For each of the fast breeder prototypes several firms were to combine in a consortium.

The Working Group received the proposal with sympathy. In particular, it was appreciated that a premature choice between the coolants would be avoided and that industrial firms would be involved in the project at a relatively early stage.

Even before the Working Group meeting, the Karlsruhe project group had initiated talks with the managements of Siemens and AEG in order to

Table 4-1
Program Schedules, 1960 to 1980

Schedule	First Phase: Preliminary Choice, Completion	Second Phase: Detailed Design and Construction of Prototype			Third Phase: Construction of Demonstration Plant		Commercial Introduction (first commercial order)
		Design Start	Construction Start	Operation Start	Construction Start	Operation Start	
1960 [a]	1963			Possibly 1970			After 1970
1962 [b]	1967		Not before 1968	Not before 1970			Not before 1975
1963 [c]	1967	1967		1972			End 1970s
1964 [d]	1967	1968		1973			1980
1964 [e]	1967	1968		1974	After 1973	End 1970s	
1964 [f]	1965–1967		1971	1970–1972	1975	1980	
1965 [g]		1965	1965–1967	1972–1973	1973	1977	Mid-1970s (abroad)
1966 [h]		1966	1968–1969	1973	1974–1975	1977–1978	1978–1980
1966 [i]			1969	1973	1974–1975		1979–1980
1968 [j]				1974			1980
1969 [k]				1975			
Dec. 1970 [l]			1970	1976			1980–1985
May 1971 [m]			1972	1977			
1972 [n]			April 1972				1990s
Mar. 1973 [o]				1978	1979		
1976 [p]				1981	1982	Late 1980s	1990s
1980 [q]				1985	1986		
Actual dates	1967	1966	April 1973				

Sources:

[a] German Atom Commission, Working Group, appendix 3, minutes of 25th session, 30 September 1960, and appendix 2, minutes of 26th session, 7 December 1960.

[b] W. Häfele and K. Wirtz, "Ansätze zur Entwicklung eines schnellen Brutreaktors," *Atomwirtschaft* 7 (1962):557–559.

c W. Häfele, "Das Project Schneller Brüter," *Atomwirtschaft* 8 (1963):206–209; W. Häfele, "Neuartige Wege naturwissenschaftlich-technischer Entwicklung," in *Die Projektwissenschaften*, Forschung und Bildung, Schriftenreihe des Bundesministers für wissenschaftliche Forschung, no. 4 (Munich: Gersbach & Sohn, 1963), pp. 17–38.

d W. Häfele et al., "The Karlsruhe Fast Breeder Project," Paper P/539, *Third United Nations Conference on the Peaceful Uses of Atomic Energy, Geneva 1964*, proceedings, p. 73.

e German Atom Commission, Working Group, appendix 1, minutes of 46th session, 17 November 1964.

f W. Häfele, "Das Projekt Schneller Brüter," in *Taschenbuch für Atomfragen 1964*, edited by W. Cartellieri et al. (Bonn: Festland Verlag, 1964), pp. 27–31.

g Kernforschungszentrum Karlsruhe, Projekt Schneller Brüter, "Ausführliche Erläuterungen zu den Anträgen zur Bereitstellung der Mittel für die Erstellung der Unterlagen zum Bau der Prototypen des Schnellen Brüters," October 1965, p. 20 (unpublished).

h W. Häfele, "Das Projekt Schneller Brüter Karlsruhe," *Atomwirtschaft* 11 (1966):293–301; see also W. Häfele and P. Engelmann, "Das Projekt Schneller Brüter," in *10 Jahre Kernforschungszentrum Karlsruhe*, edited by Gesellschaft für Kernforschung (Karlsruhe: Gesellschaft für Kernforschung, 1966), pp. 17–27.

i W. Häfele, "Schnelle Brutreaktoren, ihr Prinzip, ihre Entwicklung und ihre Rolle in einer Energiewirtschaft," in *Arbeitsgemeinschaft fur Forschung des Landes Nordrhein-Westfalen, Natur, Ingenieur- und Gesellschaftswissenschaftern*, n. 163 (Cologne and Opladen: Westdeutscher Verlag, 1966), pp. 49–74.

j W. Häfele, "Federal German Fast Breeder Programme," Report EURFNR-403, *Proceedings of the Conference on Fast Reactor Physics*, vol.1 (Vienna: International Atomic Energy Agency, 1968), pp. 3–14.

k W. Häfele, "Zur Entwicklung des natriumgekühlten Schneller Brüters," *Atomwirtschaft* 14 (1969):212–216.

l W. Häfele, in Deutscher Bundestag, Ausschuss für Bildung und Wissenschaft, "Anhörung von Sachverständigen zu dem Thema "Wachstumsorientierte Technologien und staatliche Forschungspolitik" betr. den Bereich Kernenergie, Öffentliche Informationssitzung am 17 Dezember 1970," pp. 27–18, 27–93.

m W. Häfele, "Die Entwicklung Schneller Brutreaktoren und der SNR 300," *Atomwirtschaft* 16 (1971):247–250.

n U. Däunert and W.J. Schmidt-Küster, "Das Projekt SNR-staatliche Förderung und internationale Zusammenarbeit," *Atomwirtschaft* 17 (1972):363–366.

o K. Traube, "Der SNR-300 und die internationale Situation der Schnellbrüterentwicklung," *Atomwirtschaft* 18 (1973):411–414.

p Bundesministerium für Forschung und Technologie, "Entwicklung fortgeschrittener Reaktorlinien," Deutsches Bundestag, Ausschuss für Forschung und Technologie, Ausschussdrucksache 94, 18 January 1976 (unpublished); A. Brandstetter and A.W. Eitz, "The Tripartite Fast Breeder Programme: A Utility/Industry View," *Nuclear Engineering International* (July 1976):40–43.

q Interview with participant, January 1980.

get their consent to a participation of their reactor departments in a government financed fast breeder program.[7] Under Karlsruhe's prodding, the formation of the consortia proceeded very quickly. For a discussion of this matter, a meeting of representatives from the interested firms was organized in May 1965 within the framework of the German Atomic Forum.[8] They agreed that collaboration of firms should be established on the broadest possible level. The decision of firms for either of the two fast breeder prototypes was alleviated by an informal agreement that information should be exchanged among the firms after completion of the prototype phase.[9]

The membership in the consortia was partly predetermined by the commitments of AEG to the boiling-water reactor with nuclear superheating, and of Interatom to the sodium-cooled thermal reactor. This suggested that AEG be part of the consortium for the steam-cooled fast breeder and Interatom part of the consortium for the sodium-cooled fast breeder. AEG joined forces with Gutehoffnungshütte Sterkrade AG (GHH) und Maschinenfabrik Augsburg-Nürnberg AG (MAN). The latter are both part of the Gutehoffnungshütte concern, one of the larger West German manufacturers of heavy-plant equipment. While AEG assumed the work on the reactor, GHH took over the turbine-compressor and MAN the Löffler-boiler.[10] AEG was not particularly keen on the steam-cooled fast breeder in the sense that it was given clear advantage over the sodium-cooled fast breeder, but the design work of the steam-cooled fast breeder was a welcome employment for the nuclear superheat group.[11]

For the sodium-cooled fast breeder reactor, Interatom combined with Siemens. While Interatom focused on components for sodium-cooling, such as pumps, heat exchangers, and steam generators, Siemens was in charge of the reactor core.[12] The collaboration of Interatom with Siemens seemed reasonable because Interatom did not have experience yet with construction of conventional or nuclear power plants. Furthermore, cooperation with a large partner seemed an advantage to Interatom, as this was a relatively small firm that had no business outside nuclear technology and lived nearly totally on government-sponsored work. The fast breeder project not only helped the KNK reactor over the hurdles, but also offered employment following completion of this reactor. To Siemens, cooperation with a smaller partner which had no experience in power-plant construction was not particularly attractive. Doing it alone would have been more pleasant. But the company did not buck the wishes of the Karlsruhe project management and the ministry, which wanted Interatom in the project.[13]

By mid-1965, the formation of the industrial consortia was completed and the firms submitted their applications for government funds to the Science Ministry, the successor to the Atom Ministry. The active role of the Karlsruhe project management is nicely illustrated by the fact that it supplied a paper explaining and justifying the firms' applications.[14] In line with previous statements the paper referred to the strong international competition as the main reason for pressing ahead with the prototypes. A new element, however, was introduced: recent cost studies by the Karlsruhe center showed a clear economic advantage of the fast breeder over the light-water reactor (LWR) for the 1970s (see table 4–2). A new time schedule for the program was set, envisioning design of a sodium-cooled prototype till 1968 and construction till 1972; design of a steam-cooled prototype till 1969 and construction till 1973; and completion of a 1,000-MWe demonstration plant by 1977 (see table 4–1).

The final consultations within the advisory bodies of the German Atom Commission took place between December 1965 and May 1966.[15] The Working Group set up an ad hoc committee to look at technical details and to work out proposals for financing the design work, for the organizational arrangement, and for possible collaboration with other European countries. This ad hoc committee comprised three members of the Working Group, two officials of the ministry, and a university professor who was not a member of the Working Group. It met in two sessions in December 1965

Table 4–2
Electricity Costs as Estimated by the Karlsruhe Laboratory in 1965

		Electricity Costs	
Type of Reactor		DPf/kWh	mills/kWh
Liquid-metal-cooled fast breeder reactor II (General Electric)	LMFBR II	1.62	4.05
Liquid-metal-cooled fast breeder reactor I (Karlsruhe laboratory)	LMFBR I	1.62	4.05
High-temperature gas-cooled reactor (General Atomic)	HTGCR	1.80	4.50
Light-water reactor of pressurized-water type I (ORNL)	LWR I	1.91	4.78
Advanced gas-cooled reactor (UKAEA)	AGR	2.02	5.05
Heavy-water reactor (Siemens)	HWR	2.09	5.23
Light-water reactor of pressurized-water type II (Siemens)	LWR II	2.12	5.30
Gas-graphite reactor of Magnox type (CEA)	Magnox	2.19	5.48

Sources: H. Grümm et al., "Demands for Nuclear Fuels and Costs of Different Reactor Types in Germany," Report KFK-366 (Karlsruhe: Gesellschaft für Kernforschung, September 1965), p. 60; reprinted with permission. More detailed information on estimates is given by H. Grümm et al., "Ergänzendes Material zum Bericht 'Kernbrennstoffbedarf und Kosten verschiedener Reaktortypen in Deutschland' (KFK-366)," Report KFK-466 (Karlsruhe: Gesellschaft für Kernforschung, September 1966).

and April 1966. Its considerations were reported to the Working Group in February 1966 and May 1966, which discussed and approved them. Final approvals of the Expert Commission and of the Atom Commission were obtained in February and March 1966.

After the final details had been negotiated with the firms involved and with the Ministry of Finance, the Ministry for Scientific Research issued in November 1966 the appropriation statements. DM57.5 million was assigned for the design of the sodium-cooled prototype plant to Interatom (DM35 million) and Siemens (DM22.5 million); DM38.7 million for the steam-cooled prototype plant to AEG (DM24.7 million), GHH (DM7.5 million), and MAN (DM6.5 million). The funds appropriated to the consortia were exactly those demanded. The ministry's justification for the expenditure followed the Karlsruhe arguments referring to the vast commercial potential of the fast breeders in the near future and to the pressing international competition. Unless the challenge of international competition was met, the American lead was seen to jeopardize the fruit of the whole West German effort so far.[16]

European Collaboration

The Euratom commission still had plans for a unified European fast breeder project. In early 1965 it started an initiative to achieve this unification at the prototype phase that was to follow the expiration of the association contracts in 1967. The intention was to concentrate the basic R&D at Cadarache and Karlsruhe. As soon as possible a consortium of European firms was to be established which would initiate design studies for a common prototype plant and later serve as main contractor for construction of the plant.[17]

This proposal competed with the Karlsruhe plans. Karlsruhe would have liked to see the continuation of its association with Euratom in order to have Euratom finance 40 percent of its work. But a joint European prototype would have removed program leadership from Karlsruhe. In the discussions of the Working Group the Karlsruhe scientists argued strongly against the Euratom proposal. Participation of Euratom in the two West German prototypes would have subjected these plants to a Euratom—United States information-exchange agreement. This was not regarded as desirable at a time when American industry was about to start a sales attack. Furthermore, the Karlsruhe scientists deemed a single European prototype to be ineffective in meeting American competition.

Nevertheless, the Atom Commission recommended in May 1965 the construction of a joint sodium-cooled prototype with France in order to reduce its cost to the West German government. It was understood that this would have been done outside the framework of Euratom. In the Working Group, however, the interests of Karlsruhe and the reactor groups in industry prevailed. To them participation in a French prototype, to be designed and constructed by the Commissariat à l'Energie Atomique (CEA), was not attractive. They preferred to proceed with two domestic prototypes.

However, there was no need to resolve the differences of opinion between the Atom Commission and the Working Group. The idea of a joint French-German sodium-cooled prototype plant was discarded in summer 1965. This was only partly the result of the reluctance of West German actors. At least as strong a reason was the Gaullist policy toward the European communities.

Finally, a joint prototype with France was seen to compete with the emerging collaboration of the West Germans with Belgium and the Netherlands. The attitude of the Karlsruhe group toward collaboration with Belgium and the Netherlands was more favorable, as it supplemented the West German program without questioning the leadership of the Karlsruhe laboratory. Furthermore, this collaboration recruited considerable political support for the proposals of the Karlsruhe group for work on two prototypes. The Karlsruhe project manager could report to the Working Group that industrial firms in the Netherlands were willing to cooperate with the West German program, and that the Belgian and Dutch governments were prepared to make financial contributions to the sodium-cooled prototype.

Talks of West German government officials in Belgium and the Netherlands during the first half of 1966 confirmed the willingness of the Belgian and Dutch governments to collaborate and share the costs. The attitudes of German officials toward this collaboration were determined by political motives rather than by commercial and technical aspects.[18] They reckoned that the collaboration of the Dutch and Belgians would not make the project cheaper, and that it would increase inflexibility. Nevertheless, some advantages were seen in broadening the technological base and in assuring a larger market. Among the political motives, the strongest related to the peculiar postwar situation of West Germany in which the government had to circumvent any suspicions about potential military uses of the plutonium entailed in a fast breeder program. In the mid-1960s, this problem was accentuated by discussions concerning the nonproliferation treaty. International collaboration was welcomed as a means to protect the West German nuclear industry against potential discriminatory application of this treaty. Furthermore, the government saw in this collaboration a welcome chance to document its commitment to European integration. After all, the inclination of Belgian and Dutch institutions to join forces with West Germany was taken as a confirmation of the ministry's fast breeder policy.

The West German firms involved in the fast breeder project were not enthusiastic about sharing the work with Belgian and Dutch companies. But once Karlsruhe and the ministry had taken the lead, they followed suit without much resistance.[19] West German know-how on sodium technology was ahead of that in the Netherlands. Hence this aspect of collaboration was not considered very attractive. Belgonucléaire, however, because of its contributions to the Enrico Fermi reactor in the United States, had know-how in fuel-element technology from which the Germans could profit.

Organizational Framework

The Science Ministry in general appreciated Karlsruhe's proposal for participation of industrial firms in the design of the prototypes, as it minimized criticism of the project by industry. A broad participation of industrial firms helped to spread the ministry's support evenly over West Germany's nuclear industry. With Siemens/Interatom working on the liquid-metal fast breeder, AEG/GHH/MAN on the steam-cooled fast breeder, and Brown, Boveri & Cie/Krupp on the high-temperature gas-cooled reactor (HTGCR), each of the three large electrical firms had its project in the ministry's advanced reactor activities.[20]

Furthermore, early participation of industrial firms in the fast breeder program appealed to the ministry as a cure for some flaws it saw in the Karlsruhe project.[21] Ministry officials had become aware that the work in Karlsruhe was biased toward reactor physics, and that engineering aspects were not given sufficient attention. Industry participation now opened a way to have more engineers looking at fast breeders and to strengthen the engineering side in the project. For the reactor section of the ministry, a further attraction of the involvement of industrial firms was tighter control of the fast breeder project. To officials of this section the fast breeder activities in Karlsruhe appeared to be somewhat uncontrolled. As the Karlsruhe project management had very close ties with the top officials in the ministry, the reactor section itself had little say in the project. Industrial participation now offered a level to curb the influence of the Karlsruhe project management and to bring a clear line into the fast breeder activities of the laboratory.

The organizational framework for the design and construction of the two prototypes was therefore to become a matter of dissent between the Karlsruhe laboratory and the ministry's reactor section. The Karlsruhe project management proposed that it give the design contracts to industry, and that the order for construction of the prototypes then be placed later by a company founded jointly by government research laboratories and utility companies of the three countries. After completion the prototype plants were to be transferred to the utilities.[22] Officials in the ministry's reactor section, however, wanted the ministry to give the design contracts. They wanted to subordinate the Karlsruhe project management to industrial activities already in the design phase, and even more in the construction phase.

In the final arrangement, the ministry's reactor section prevailed. The design contracts were commissioned to industry by the ministry. Karlsruhe was obliged to enter collaboration agreements with the industrial consortia, which were to specify the information that Karlsruhe would supply for their design work. According to the ministry's intention, industry as the customer of the Karlsruhe laboratory should determine in these agreements what kind of research results it wanted from the laboratory. This conflict of interest dragged on into the negotiations of these collaboration agreements, causing some dealy. Other means for the ministry to get a closer control on the Karlsruhe project were seen in systems analysis and planning techniques, which were fashionable at that time. Hence the ministry urged the Karlsruhe project to deploy a network planning technique for its work.

In a special appendix to the appropriation statement of November 1966 the ministry defined the results in component and irradiation tests that were to be attained before a decision on the construction of the two prototype plants was to be taken. This bureaucratic innovation also aimed at a tighter control over the project, especially at preventing pressures from the Karlsruhe laboratory for a premature construction start.

Finally, the Karlsruhe influence was reduced by the establishment of a new governing committee for the project: the Fast Breeder Project Committee, comprising four representatives from the consortia, two from the Karlsruhe laboratory, two from utilities, one from a nuclear fuel company, and two civil servants from the ministry. The committee met about twice a year. However, it did not exert a strong leadership. Responsibilities were not clearly separated between Karlsruhe and industry in subsequent years, and the Karlsruhe project management claimed direction not only of Karlsruhe's own work but also of industrial activities.

Costs and Finance

In its October 1965 paper explaining the firms' applications to the ministry, the Karlsruhe project management estimated the costs of the sodium-cooled prototype at DM310 million and of the steam-cooled prototype at DM335 million. The former was assumed to be less expensive than the latter, as the Karlsruhe scientists expected that sodium components originally fabricated for testing could later be installed in the prototype plant. In addition to the construction costs, funds were required for design, engineering, and component testing: DM170 million for the sodium-cooled prototype and DM50 million for the steam-cooled prototype.

For the work at Karlsruhe over the period 1968 to 1972 costs were estimated at DM170 million. Altogether the cost of the seond phase of the project was estimated at DM1,035 million. This sum included all expenses up to completion of the prototypes. For the design and construction of the 1,000-MWe demonstration plant no government subsidy was thought necessary, except for a government guarantee covering possible losses in operation resulting from technical failures. Including an estimated total of DM356.1 million for the first phase of the project, the whole project from its start up to commercialization was estimated to cost DM1,391.1 million. A good share of this sum was to be financed by the Belgian and Dutch governments and by Euratom. The utilities' contribution to prototype construction was assumed at DM150 million for each reactor. The West German government would thus have to finance DM149.5 million for the first phase, and DM346 million over the second phase (1966–1973), or an annual expenditure of DM43.25 million for phase two.

The Karlsruhe scientists compared this expenditure with the benefits to be derived from the FBR. According to their calculations, commerical application of fast breeders would produce a cumulative net saving of about DM1,000 million by 1984. This meant that by then the fast breeder program would have produced a net earning for the national economy.

The committees of the Atom Commission showed considerable restraint in dealing with these estimates. The Working Group accepted in December 1965 the Karlsruhe paper as a guideline. Only one ministry official participating in the ad hoc committee set up by the Working Group felt that the cost estimates for prototype construction were too optimistic. Construction costs were therefore set at DM450 million for each prototype, raising total expenditure for the second phase from DM1,035 million to about DM1,300 million. This figure was later accepted by the Working Group and other committees of the Atom Commission without discussion.

The question of a financial contribution by industry to prototype design received more attention and was raised several times in the decision making. However, the minutes of the advisory committees show that the Science Ministry did not put much pressure on firms. In the first report on the Karlsruhe proposal to the Atom Commission by the chairman of the Expert Commission III, a financial contribution of industry to the design work was excluded because of high cost and risk and because prospects of success were not yet assessable.

When the Working Group set up in December 1965 the ad hoc committee to look, among other things, on possible schemes of finance for the prototype work, a senior official of the ministry stated beforehand that it would be wrong to base the finance of prototype design on the assistance of utilities or manufacturers. A utility representative informed the Working Group that the utilities thought it premature to expect them to make a financial contribution toward prototype work. Hence it was not much more than a formal exercise, when the ministry raised the question of financial participation anew in letters to the firms in early 1966.

At the subsequent meeting of the Working Group a speaker of manufacturing industry pointed to big uncertainties involved in the project and stated that a financial contribution from the supplier side could not be expected as long as an economic utilization of FBRs was not foreseeable. The Working Group finally recommended that the ministry should not demand a financial contribution from firms doing design work for the prototypes.

These statements clearly show that manufacturing firms and utilities did not share Karlsruhe's optimism about the fast breeder's economic potential. It was not the scale of the funds required but the uncertain economic prospects of the fast breeder which made the firms unwilling to invest their own money. The DM96 million allocated by the ministry in November 1966 represented an annual expenditure of DM32 million. This is 0.2 percent of the combined sales of AEG, Siemens, and the GHH group in 1964–1965, or 0.1 percent of the combined sales of these firms and the parent companies of Interatom in the same year.[23] In 1967, the West German electrical industry spent an average of 6.3 percent of sales on R&D.[24] Assuming this research intensity for these firms, the design of prototypes would ·have taken about 3 percent of the firms R&D budget if only AEG, Siemens, and GHH were considered, or about 2 percent if the parent companies of Interatom were included. This shows that, if the manufacturers could have been confident in supplying a commercial FBR by 1974, the expenses for the development of prototypes would not have been beyond their financial reach.

Economic Assessment

Pressing ahead with the fast breeder appeared necessary to the Karlsruhe project management for two reasons: strong international competition, and prospects in the near future for electricity costs in FBRs underbidding those in LWRs. Both points were taken as signs that there was a vast commercial potential for fast breeders long before the depletion of low-cost uranium resources had an impact on the economics of nuclear power. This difference between long-term and short-term prospects of the breeder reactor was underlined in Karlsruhe's justification for an acceleration of the program, stating that acceleration was not warranted by the long-term potential of fast breeders, but only by their near-term commercial prospects as expected for the 1970s.[25]

The Science Ministry accepted the Karlsruhe assessment with only marginal reservation.[26] As mentioned earlier, an official did not accept the Karlsruhe estimates of construction costs of the two prototypes, and proposed a somewhat higher figure. But Karlsruhe's perception of an imminent international competition and its estimates of future electricity costs were taken for granted uncritically by the ministry. It had little incentive for further scrutiny. The bright economic prospects of FBRs as indicated by the coming American competition and by the Karlsruhe electricity-cost estimates seemed to confirm prior policy decisions. The emerging collaboration with Belgium and the Netherlands implied a gain in reputation not only for Karlsruhe but also for the ministry. Earlier doubts about the ministry's role in the development of nuclear reactors were dispersed. The prospect of West Germany catching up in civilian power with other Western countries appeared realistic. Top officials indeed stated with pride, that the commitment to sodium-cooled and steam-cooled fast breeder prototypes, together with a commitment to an HTGCR prototype, made the West German reactor program the most ambitious in the world.[27]

Reactor manufacturers and utilities did not share Karlsruhe's optimism. As mentioned earlier, their disinclination to put their own money into prototype design revealed an economic assessment that was in blatant contradiction to the image of an imminent sales attack and of splendid near-term economic prospects as pictured by the Karlsruhe project management. As interviews with a number of participants confirmed, the larger reactor manufacturers and the utilities did not consider the reports about GE's fast breeder plans to be worth particular attention; and they regarded the Karlsruhe cost projections as a numbers game that was not to be taken seriously.[28] In the policymaking, however, the firms' skepticism surfaced only when a financial contribution was discussed. As the basic document giving a detailed explanation and justification for the proposed decisions was supplied not by industry but by the Karlsruhe project management, the firms participating in the prototype design work were not urged to identify themselves with the Karlsruhe assessment.

At no point of the decision making were the Karlsruhe electricity-cost estimates and the perceptions of an imminent sales attack discussed in length and detail worth a mention in the minutes of the advisory bodies. As the ministry appeared receptive to earlier industrial criticism about a government laboratory engaging in design and engineering activities and now valued industrial participation very highly, detailed questions about the viability of the Karlsruhe assessment would have been considered as pedantry by industrial representatives. After all, at least for Interatom and for the nuclear superheat group of AEG the design contracts for the fast breeder prototypes provided welcome employment.

Experience of later years has proved the Karlsruhe laboratory's estimates of electricity costs to be grossly off the mark. Fears about American competition turned out to be unfounded, and the more skeptical assessment of industry warranted. The year 1974 has passed without a commercial FBR anywhere in the world. At present, official statements expect industrial use of FBRs around the year 2000, and there are still great uncertainties as to whether these expectations will pan out (for more details see chapter 7).

Why was there such a difference between the views of Karlsruhe and industry? Was it a matter of chance that industry placed the better bets? A more detailed analysis of the Karslruhe assessments reveals that both the laboratory's remoteness from the commercial world and the tactics of political decision making played a role.

Spurious Competition

What the Karlsruhe project management perceived as an imminent sales attack by GE was in reality a skirmish between this company and the U.S. Atomic Energy Commission (AEC) over the commission's advanced reactor program. Whereas AEC wanted to proceed with both a near-term program for advanced converters and a longer-term program for fast breeders, GE would have liked to see the commission skip the advanced converter program in favor of a priority commitment to FBRs.

In its November 1962 report to President Kennedy the AEC had outlined a new program for reactor development.[29] As LWRs were on the threshold of economic competitiveness, the commission's activity was to focus on a near-term program for the development of advanced converter reactors and on a longer-term program for FBRs. The commission expected FBRs to become economically attractive in the 1980s. As many decades would pass before FBRs replaced thermal reactors, advanced converters were seen to have a significant economic role. In concrete terms, the new program envisaged construction of seven to eight prototype reactors over the next twelve years, mainly financed with government money. Half were to be advanced converters and half breeder reactors.

Although the development of advanced converters was justified by their potential for a better fuel utilization, the AEC's program had more to do with industrial policy than with fuel economy.[30] The commission's postwar effort in civilian nuclear power had resulted in GE and Westinghouse establishing a strong market leadership with their LWRs, the rest of industry including companies such as General Atomic and Atomics International being left behind. Obliged by law to establish a competitive nuclear power industry, the commission hoped to help the lagging members of the nuclear industry into commercial business through government support for the construction of advanced converter prototypes. Thus the 1962 report mentioned as advanced converters the spectral–shift–control reactor of Babcock & Wilcox, the HTGCR of General Atomic, the sodium-graphite and the organic-cooled reactors of Atomics International, and the heavy-water reactor designed then by DuPont's Savannah River Laboratory. Later the light-water breeder reactor designed by the AEC's naval reactor branch was added to the list.[31]

Westinghouse and GE challenged publicly the program's justification. They argued that the advanced converters proposed by the AEC would not catch up with LWRs, and that advanced versions of their own light-water designs offered at least the same potential as any other advanced converter.[32]

The AEC nevertheless continued with its advanced reactor program and issued in March 1964 invitations for a round of five advanced converter prototypes to be constructed with government support.[33] In canvassing for this program the commission even went as far as to speak of the possibility of a "power reactor gap" between the present LWRs and the advent of the FBRs.[34] This concurred with the long-term stance the commission took on the fast breeder. Its schedule as outlined in December 1963 envisaged construction of a 200-MWe prototype plant from 1969 to 1973. Two to four demonstration plants of larger size were to follow in the early 1980s, and the final target was operation of a commercially competitive 1,000-MWe plant by 1989.[35]

This is the context in which GE announced in September 1964 its confidence in being able to offer a commercial breeder by 1974. The statement was meant to contest the AEC's justification for the advanced converter program. This program was of benefit only to GE's would-be competitors and reduced the funds available for the AEC's fast breeder program and thus for GE's own fast breeder group. If FBRs could be made available within a decade, there was no need for advanced converters. The statement was made exactly at the time when the AEC went into budgetary negotiations for the three prototype plants it had selected for its advanced converter program.[36]

For a regular reader of the technical press neither the advanced converter issue nor the time schedule proposed by GE was a surprise. When seeking utility support for its fast breeder work in 1962, GE had indicated that construction of an economically competitive, large FBR could begin in 1975.[37] In 1963 a GE manager had underlined his critique of the AEC's advanced converter program by saying that with a well-designed development program the FBR "could achieve commercial use by 1970."[38]

The September 1964 statement that made so much impression on the Karlsruhe scientists referred to a "heavy company financial investment in the breeder program" that in the view of GE officials "imposes an obligation to try to shorten the development period projected by AEC."[39] Could this be taken as an indication that GE was willing to put considerable amounts of its own funds in the development of FBRs? In the early 1960s GE was in fact the only American manufacturer entertaining a significant effort on FBRs.[40] However, this effort was mainly financed with government and utility money. Work had begun at the end of the 1950s, when the company was pretty confident that its boiling-water-type LWR would make the step to competitive nuclear power in the mid-1960s.

AEC expenditures for GE's fast breeder work totaled $2.15 million up to 1963. The major focus of subsequent work was an experimental reactor that later became known as SEFOR. Although encouraged by AEC to submit a proposal, GE failed in 1962 to get AEC funding for this reactor. A year later, however, the Southwest Atomic Energy Associates (SAEA), a group of fifteen utilities, selected the SEFOR proposal in a competition it had invited for a development contract. The financial contribution of the utility together with the contribution from Karlsruhe finally prompted the AEC to provide government funds. The SEFOR contract signed in May 1964 estimated the cost of the reactor at $12.35 million. SAEA was to contribute $5.9 million, Karlsruhe and Euratom $5 million. GE obliged itself to pay $1.45 million from its own funds and to assume all risks of cost overruns.[41] The AEC put up $12.9 million for fuel fabrication, training, and associated R&D work.

It is this commitment to which GE company officials referred to in their September 1964 statement as "the heavy company financial investment in the breeder program," which "imposes an obligation to try to shorten the development period projected by AEC." For the regular reader of the technical press this was not a new reflection. When the SEFOR contract was notified, GE thought that a successful series of tests in this reactor, together with irradiation tests in facilities planned by the AEC, would provide the technical information required for a larger plant. Thus a larger plant could be ventured in advance of AEC's accepted schedule.[42]

The goal of a commerical FBR by 1974, as announced in September 1964, did not imply the company's willingness to increase commitment of its own funds. GE hoped to round up utility support for the design of a prototype plant, that then was to be constructed within the AEC's program. Since 1965 the company, *Nucleonics Week* reported, "was out beating the bushes for utility sponsors of a fast breeder development program."[43]

This effort was again motivated by a controversy over the AEC's reactor program, now about the role of industry in the AEC's fast breeder program rather than about advanced converters. Under the direction of Milton Shaw, who became in December 1964 head of AEC's Division of Reactor Development and Technology, the commission set up a new fast breeder program that emphasized thorough component development and testing to be performed in government laboratories, before funds would be given to industry for design and construction of prototypes. GE's fast breeder group sought a greater involvement in AEC's activities and a greater share of its funds. Utility support was to be a lever to put pressure on the AEC for an early start of prototype work.[44]

By the time the Karlsruhe laboratory had the impression of an impending American sales attack, GE had not started a detailed design of a larger fast-breeder prototype plant. Apart from the SEFOR reactor, the company's design work on FBRs was of a preliminary character up to 1967. In that year, the company managed to get $5 million from the ESADA utility group and some funds from other utilities for design work on a 300-MW sodium-cooled fast prototype reactor. The three-year study sponsored by ESADA had the objective "to prepare a proposal to the AEC for the cooperative construction of a 300-Mw demonstration plant somewhere in New York State on a 1975 completion schedule."[45] In the same year a $1.2 million design and feasibility study of an experimental steam-cooled fast reactor, sponsored by a different utility group, was begun.[46]

As Karl Cohen, head of GE's Advanced Products Operation, pointed out in 1966 to a congressional committee, a utility-sponsored design study would not imply a commitment to construct a prototype. This would be contingent upon progress in the AEC's fast breeder program, favorable results of SEFOR, confirmation of earlier cost estimates, and continuation of a high growth of the nuclear power industry.[47]

From this account it is apparent that GE in the 1960s engaged in fast breeder activities of a notable scale only when funds from outside sources were at hand. Its fast breeder effort appears to have been much smaller than that of the Karlsruhe project. As of 1965 GE's Advanced Products Operation, which in addition to FBRs was also working on reactors for desalination, naval propulsion, and space applications, comprised 100 scientists and technicians. The Karlsruhe group numbered in 1966 400 scientists and technicians, with an additional 300 to 400 being involved indirectly.[48]

Westinghouse, which the Karlsruhe scientists regarded as a second strong competitor for the West German fast breeder program, had done some minor studies with AEC financing in the early 1960s. Continuous work on FBRs began in 1966, when the company announced that it was ready to commit $10 million toward preparation of a prototype design. Another $40 million were thought to be required for this work from utilities and the AEC.[49]

With the commercial breakthrough of LWRs in 1963–1965, both GE and Westinghouse made large investments in this technology. They could not hope to retrieve these investments before a decade, so they had little incentive to put additional finance in a risky effort to market a breeder reactor by the mid-1970s. An article in the March 1967 issue of *Fortune* states the wait-and-see attitude of the companies very clearly:[a]

> The attention and resources of G.E. and Westinghouse—and, to a lesser extent, Babcock & Wilcox and Combustion Engineering—have been

largely usurped by mammoth orders for light-water reactors. . . . G.E., for one, doesn't foresee any need at all for advanced converters and is not pressing for quicker development of breeders. On the other hand, General Atomic and Atomics International have little hope of selling light-water reactors and would have much to gain from the introduction of a new generation. One G.E. official is convinced that controversy over the speed at which breeder development ought to be pressed is a direct outcome of the present competitive situation. He says, "The only panic I see is on the part of people out of the business who have to be in. And for them, any flag is good."[50]

Looking at the information on GE's fast breeder activity as available in the technical press, it appears quite understandable that West German manufacturers and utilities paid little attention to announcements by the company about early offers of a commercial FBR. Why then was the Karlsruhe project management so much impressed? Two answers are possible: either it was mistaken, or it used the American competition as a convenient argument for getting the German program over the next decision point. Perhaps a mix of both answers fits reality best. As a matter of fact, the scientists' sense of a pressing international competition vanished soon after the government decision had taken the major hurdles. Whereas in 1965 the need for accelerating the project had been communicated to the press, statements by the Karlsruhe scientists in early 1966 no longer display any sense of urgency.[51]

Evidence of Estimates

From empirical research on technological innovation it is known that estimates of the cost of a new product are uncertain. Cost-overruns are the rule rather than the exception.[52] In a study on twenty-two technological innovations in the U.S. Air Force, Marshall and Meckling found a mean ratio of first to latest estimates of production costs of 2.4 or 3.2, depending on what kind of adjustments were made.[53] For projects incorporating a large technological advance the ratio was considerably higher (3.4) than for those incorporating a small technological advance (1.3 or 1.4). Data from military projects may not, however, be representative for civilian projects, as the military usually gives time and performance targets a higher priority than cost targets, and as there is a trade-off between these targets.[54] Available evidence seems to confirm the findings of Marshall and Meckling for the civilian sector. In a study of seven chemical innovations, Allen and November noted considerable inaccuracies in production-cost estimates, with an optimistic bias.[55] In a study of ten electronic instruments, Thomas found that cost estimates tended to be accurate if a small technological advance is sought, but that for products incorporating a medium or large technological advance the costs tended to be underestimated by a mean margin of about 60 percent.[56] For new types of power-generating plants such as the fast breeder, it follows from such general considerations that they belong to a particularly risky class of innovations. FBRs represent a large advance in technology compared to proven types of nuclear power plant, and thus one can expect large inaccuracies in forecasting costs.

Interviews with individuals in the manufacturing and utility industries showed that even in activities other than R&D the uncertainty of cost estimates is part of everyday business life.[57] It is common knowledge that firms quote optimistic prices as long as they are not obliged to supply at the quoted price; that firms give optimistic design specifications as long as they do not have to give a guarantee; and that a price means very little as long as an order is not clearly defined. In the power-plant business these uncertainties are particularly great. Utilities generally regard price estimates by a supplier as mere speculation, as long as the design of a plant is not detailed enough to enter contract negotiations. As far as new types of plants are concerned, they are very sensitive about the risk of a plant not fulfilling its design specifications. Hence the critical design specifications usually are not taken for granted before a supplier is prepared to give a guarantee.[58]

The attitude of individuals in the reactor manufacturing industry can be grasped from the following statement of the head of reactor development at AEG in 1971:

> I want to clear away the conception that with regard to a reactor which has yet to be developed costs [of electricity] can already be calculated to the first or second decimal. In developing light-water reactors, we have had the experience that at least ten years of development were required before the order or magnitude of costs could be assessed to some extent and before further development became routine as in any product, so extrapolation was possible and future costs could be grasped.[59]

From hindsight this statement may even appear too optimistic, as it was made before AEG incurred the large losses that later caused the company to quit the nuclear power business.

In presenting their electricity-cost estimates to the Science Ministry, the Karlsruhe scientists characterized these not as uncertain guesses but as clear evidence. Their estimates, they said, showed very clearly that according to present knowledge the FBR had the greatest economic potential. Although they recognized that some uncertainty was involved, they stressed that their figures allowed almost no doubt, pointing out that their calculations did not take into account the future development potential of the FBR.[60]

Why did the Karlsruhe scientists see so little uncertainty in their estimates? The answer lies partly in the inexperience of the individuals who made these estimates, partly in a liberal use of rhetoric by those who proposed them to the government and to the public.

The estimates were made by a study group on nuclear energy reserves at the Institute of Applied Reactor Physics. This group was assembled in fall 1964 and published its report in September 1965.[61] It consisted of young scientists, mainly physicists. Most of them had come to the Karlsruhe laboratory directly from university and had no experience in industry or business. There was no economist in the group.[62]

The scientists surveyed all economic data on different reactor types available in the open literature. Additional estimates were solicited from West German reactor manufacturers and from some organizations abroad. Among this vast amount of data, eight designs of nuclear power plants with six different types of reactors were chosen for comparison. Only units of 1,000 MWe were selected, as this rating was expected to be common in the future and allowed a consistent and convenient comparison. Since published information on nuclear power economics is not simply comparable due to differences in the definition of cost items and to differing assumptions on tax and interest rates, the scientists developed a set of consistent cost definitions and economic ground rules.

The final comparison as reproduced in table 4–2 comprised two estimates for FBRs that both arrived at 1.62 DPf/kWh (4.05 mills per kilowatt-hour). One was for a GE design (LMFBR II): its cost data were taken from a GE report published in January 1964.[63] The other (LMFBR I) was a Karlsruhe design that had been completed in December 1964.[64] Estimates for Magnox and advanced gas-cooled reactors (AGRs) were based on direct communication with the French Commissariat à l'Energie Atomique and the British Atomic Energy Authority, respectively. Information from Siemens was used for the estimates for a heavy-water reactor and a light-water reactor (LWR II). A report of the Oak Ridge National Laboratory (ORNL) was the source for a light-water reactor (LWR I) and for a HTGCR of General Atomic.[65]

Thus only one of the cost estimates was based on design work performed at the Karlsruhe laboratory. But even this study, which was done at the Institute of Neutron Physics and Reactor Technology, relied for its cost estimates mainly on the literature. The fuel-cycle cost data were taken from a GE report.[66] For their capital-cost estimate, the scientists asked manufacturing firms and utilities to give the price of specific components. The addition of component estimates left considerable scope for contingencies. These were finally determined so as to put the cost of a liquid-metal fast breeder a little higher than that for an LWR as given in the technical literature. Thus even the capital-cost estimate for the Karlsruhe FBR design leaned on published views about future LWR costs.[67]

Following a January 1965 report from the ORNL with lower capital-cost estimates for LWRs than hitherto published,[68] the scientists in the study group on nuclear energy reserves felt justified in reducing the earlier capital-cost estimates their colleagues had made on the FBR from DM462.6 million to DM415.4 million (see table 4–3). As these are direct capital costs, such a change has nothing to do with adjustments of cost definitions or of economic ground rules. This change simply documents the extent to which the Karlsruhe scientists, when estimating the capital costs of the Karlsruhe design for an FBR, took guidance from published information on LWR.

The great trust which the scientists put on published cost information was the result of their inexperience. Their optimistic assessment of FBRs was derived from the views of fast breeder advocates abroad, in particular the fast breeder group of GE. Due to their inexperience, the Karlsruhe scientists regarded the GE estimates as more reliable than did the company itself. As for reprocessing, GE's estimate was based on specifications by the AEC for a reprocessing plant, so it did not necessarily present the company's own view. The cost of major components of the nuclear-steam-supply system was estimated by dollars per pound of steel used. As mentioned earlier, GE management felt these cost estimates needed confirmation by a detailed design before construction of a prototype plant was warranted. The Karls-ruhe scientists, by contrast, regarded them as sufficient for justifying an acceleration of the German fast breeder program.

Although their cost calculations may have added to the scientists' confidence in the fast breeder's bright prospects, the scientists also saw hard and precise figures as an appropriate means to get their assessment across to the policymakers. For example, that the electricity-cost estimates for the two FBRs were identical up to the last digit (see table 4–2) was not the contingent result of the input data. If the cost data used for the calculations are compared with the source data on which they are allegedly based (see table 4–3), one can see that a number of changes were made in the fuel-cycle data and in the capital-cost data. As not all these changes can be explained by adjustments for consistent definitions and ground rules, one must conclude that input data were tuned so as to produce the same electricity costs for the two fast breeder designs. Thus the figures were capable of creating an impression as if estimates by Karlsruhe and GE coincided for different reactor designs.

Argumentative rhetoric was involved also in presenting a 15-percent advantage of the FBR over the LWR as evidence for the fast breeder's great economic potential. This is suggested by some observations that can be made on the estimates. For example, the estimates included two LWRs, one named ORNL design, the other a Siemens design (see table 4–2). According to the ORNL report the design and cost data for their reactor were supplied by Westinghouse, so both were pressurized-water reactors (PWRs). As the Siemens design for some components was based on Westinghouse licenses, the two designs were not very different. Hence the 0.2 DPf/kWh difference between both designs can hardly be taken as significant. How then could a 0.3 DPf/kWh difference between the fast breeder and the Westinghouse PWR matter so much?

A later report by the study group on nuclear energy reserves included an AEG design of an LWR plant of boiling-water type. Because of an increased burnup, this plant arrived at a cost advantage of 0.33 DPf/kWh over the Siemens design. Whereas in the 1965 publication such a difference was presented as evidence for the commercial superiority of the FBR over the LWR, the 1966 report did not hesitate to deny that such a difference mattered in assessing different types of LWRs.[69]

It is not the task of this study to ask whether the Karlsruhe scientists were legitimate in their rhetorical use of overly precise figures. Rhetoric is part of everyday political life. More important is the question as to the quality of a policymaking process that is receptive to the unrealistic assessments that underlay this rhetoric.

Tight Schedule

Although R&D activities in Karlsruhe proceeded roughly on schedule, they were in 1966 still far from a stage that could have provided a solid basis for a prototype design. Of the facilities for experimental reactor physics, the fast-thermal STARK research reactor went into operation in June 1964, the subcritical assembly SUAK in November 1964. These two facilities, however, were designed mainly for gathering basic nuclear data and for developing measurement methods for use in the zero-power critical assembly SNEAK. The SNEAK facility, the biggest and the most important of the research facilities, went into operation in December 1966. First measurements with plutonium fuel were made in 1968. This meant that the significant results of the SNEAK experiments would be available only in the final stage of the prototype design. Construction of the SEFOR reactor had started in October 1965; completion was estimated at three years, and a further three to five years were reckoned for carrying out the contracted experimental program. This meant that only the first experimental data would be available before construction start of the two prototypes, while the results of the important dynamical experiments in the SEFOR reactor would become available at best during construction.[70]

Work on fuel fabrication was up to 1965 restricted to uranium. In the absence of plutonium, the fabrication of mixed-oxide fuel was simulated by using a mixture of uranium and cerium oxide. Work on plutonium fuel started in 1965, when the plutonium from the United States arrived at Karlsruhe for use in the SNEAK facility. Irradiation experiments were

Table 4-3
Input Data for the Karlsruhe Electricity-Cost Estimates

	1,000-MWe LMFBR Karlsruhe Design		1,000-MWe LMFBR General Electric Design		1,000-MWe LWR ORNL Design
	Estimate by Karlsruhe Dec. 1964	As Adjusted by Karlsruhe Sept. 1965	Estimate by General Electric Jan. 1964[c]	As Adjusted by Karlsruhe Sept. 1965	As Adjusted by Karlsruhe Sept. 1965
Specific investment costs (DM/kW)					
Direct	462.6	415.4	(379.5)[f]	440.0	369.2
Indirect	70.9	124.6	n.a.	132.0	110.8
Interest during construction	107.3	59.4	n.a.	62.9	52.8
Total	640.8	599.4	500.24	634.9	532.8
Plant life (years)	15	25	n.a.	25	25
Annual fixed charge (percent)	13.9[a]	11.76[a]	14.26[a]	11.76[a]	11.76[a]
Fuel fabrication (DM/kg)					
Core	1,000	740	1,192	n.a.	250
Axial blanket	240	80	164	n.a.	—
Radial blanket	200	190	300	n.a.	—
Core and blanket	381[a]	300	476	476	—
Reprocessing (DM/kg)					
Core and axial blanket	330	n.a.	n.a.	n.a.	—
Radial blanket	200	n.a.	208	n.a.	—
Core and blanket	252	260	272	120	100
Spent fuel transportation (DM/kg)	0[d]	100	—	40	40

Plutonium (DM/g)	26	40	26.4	40	40
Uranium (DM/kg)[e]	0	12	20	12	70.48
Burnup (MWd/t)					
Core	100,000	79,000	110,000	110,000	21,000
Core and blanket	29,000[a]	22,860	40,000	39,000	—
Electricity costs[b] (DPf/kWh)					
Capital cost component	1.453[a]	1.150	1.16	1.218	1.022
Operation and maintenance	0.198[a]	0.119	0.12	0.119	0.119
Fule-cycle costs	0.319[a]	0.349	0.27	0.278	0.766
Total electricity costs	1.9691	1.618	1.55	1.615	1.907

Sources: D. Smidt et al., "Referenzstudie für den 1 000 MWe Natriumgekühlten Schnellen Brutreaktor (Na 1)," Report KFK-299 (Karlsruhe: Gesellschaft für Kernforschung, December 1964); H. Grümm et al., "Demands for Nuclear Fuels and Costs of Different Reactor Types in Germany," Report KFK-366 (Karlsruhe: Gesellschaft für Kernforschung, September 1965); H. Grümm et al., "Ergänzendes Material zum Bericht 'Kernbrennstoffbedarf und Kosten vershiedener Reaktortypen in Deutschland' (KFK-366)," Report KFK-466 (Karlsruhe: Gesellschaft für Kernforschung, September 1966); General Electric Company, "Liquid Metal Fast Breeder Reactor Design Study," Report GEAP-4418 (General Electric, January 1964.

[a] Figure is calculated by the author using input parameters and method of the respective source.

[b] All estimates are for a load factor of 0.7.

[c] For better comparability, the case with private owrnership of fuel was selected and the nuclear insurance (355,000 $p.a.) was included in the capital-cost component rather than in operation and maintenance. U.S. dollars were converted into deutschmarks using an exchange rate of 1 U.S. dollar = 4DM. This exchange rate was also used in the Karlsruhe estimates presented in this table.

[d] The 1964 Karlsruhe estimate assumed that reprocessing and refabrication facilities were adjacent to the reactor site, so transport costs were negligible.

[e] Waste uranium for fast breeder reactors, natural uranium for light-water reactors.

[f] The General Electric estimate breaks down investment costs into direct and indirect only for the nuclear-steam-supply system. The figure given in this table assumes that the same ratio of direct to indirect costs applies to the whole plant.

limited to samples of cladding material and of uranium and cerium oxide. However, evaluation was confined to cladding material, as the hot cells required for investigation of irradiated fuel were not completed before 1966. A number of fuel rods with plutonium oxide were prepared for irradiation in the Enrico Fermi reactor in the United States, but the Fermi accident in October 1966 negated these plans. Fuel elements for a steam-cooled FBR had not yet been tested anywhere in the world, when West Germany decided to design a large prototype. Tests were planned to be done in the HDR reactor, after refitting with a fast-thermal core. By 1965, first operation of the HDR reactor with a fast-thermal core was planned for the end of 1971. This implied that the results of irradiation tests would be available about the time the steam-cooled prototype would start operation.

No experience existed on reprocessing of fast breeder fuel. Planning for a small test plant with a capacity of 1 kilogram per day had started in 1965. This plant had its first high-active operation in 1971 and reprocessed the first plutonium fuel in 1979.[71]

As a consequence of its acceleration, the West German program as of 1966 had a very tight schedule: prototype design proceeded in parallel to experimental reactor physics and to design and construction test reactors for fuel-element irradiation. There was very little flexibility for prototype design to adjust to unforeseen experimental results. Because of limited experience in component testing before prototype construction, great technical uncertainties were to be carried over into the prototype phase. Increased technical uncertainty was the price to pay for the acceleration of the program.

This uncertainty was increased further by the fact that, while the overall fast breeder program was accelerated, not much of a hurry was felt in the HDR and KNK projects. The preceding chapter discussed the factors that prevented the KNK project from being brought in line with the requirements of a sequential, step-by-step development strategy for the fast breeder program. Also for the HDR reactor, the idea was discussed by the advisory bodies to start with a fast or fast-thermal core instead of a thermal core as originally designed. Soon after construction start in 1965 the supplier lost interest in this reactor design, as earlier fears about wet-steam turbines as used in normal boiling-water reactors proved unfounded. Nevertheless, the advisory bodies recommended sticking to an already agreed construction schedule and starting the operation first with a thermal core.[72]

Like the details of economic need, the high technical risks resulting from an accelerated program schedule were not discussed in the Working Group or other committees of the German Atom Commission. It seems that, for the advisory bodies, those details were much less important than the fact that industry was to participate in the fast breeder program and that a premature choice between the different types of FBRs was avoided. From the point of view of the development strategy, both objectives were in fact very positive. As experience in other countries such as Britain has shown, considerable problems can emerge in the commercialization of a new type of nuclear power plant if industry gets involved at a late stage and if it is given too little responsibility. Deferment of the choice between the sodium-cooled and the steam-cooled fast breeder type was warranted, as knowledge on these two types was limited and very little experience existed in component performance and fuel-element behavior under irradiation. However, the objectives of avoiding a choice between the sodium-cooled and the steam-cooled types and of getting industry involved would not have required hurrying up the prototype design. They could also have been fulfilled, at that stage of the project, by contracting to the firms the design of fast or fast-thermal cores for the HDR and KNK reactor and associated development work. It would have been sufficient to let industry go on with the two small test reactors and to see then, whether results of that work would allow a reasonable choice. This would have been a better decision than a premature start of prototype design.

Notes

1. Interviews with participants, November 1975, December 1975. The GCFBR was back on the stage again in 1967. The prestressed-concrete reactor vessel had been developed in Britain, and prospects for an appropriate fuel element had improved considerably through the conception of roughened fuel-rod surfaces and the development of vanadium alloys. Furthermore, large gas turbines seemed now feasible, which would permit a direct cycle for cooling omitting the steam generators. See K. Wirtz, in *3rd Foratom Congress London 24–26 April 1967: Industrial Aspects of a Fast Breeder Reactor Program* (London: British Nuclear Forum, 1967), pp. 26–27; idem, "Zur Entwicklung gasgekühlter Schneller Brüter," *Atomwirtschaft* 14 (1969):216–218.

2. W. Häfele et al., "The Karlsruhe Fast Breeder Project," Paper P/539, *Third United Nations Conference on the Peaceful Uses of Atomic Energy, Geneva 1964*, proceedings, p. 75.

3. German Atom Commission, Working Group II-III/1 on Nuclear Reactors (henceforth Working Group), appendix 1, minutes of 46th session, 17 November 1964 (unpublished). For published information on this version of the Karlsruhe development strategy see D. Smidt et al., "Referenzstudie für den 1 000 MWe natriumgekühlten Schnellen Brutreaktor (Na1)," Report KFK-299 (Karlsruhe: Gesellschaft für Kernforschung, December 1964), pp. 1–1 to 1–4.

4. The announcement was made on 24 September 1964 at a press conference by the company presenting a price list of BWR plants; see "Atom

Power Plant Priced in New G.E. Catalogue," *The New York Times* (25 September 1964):57; "GE Issues Nuclear Plant Price List," *Nucleonics Week* (1 October 1964):2–4; "GE: Commercial Breeder by '74," *Nucleonics* (November 1964):18; "GE Publishes Prices," *Nuclear News* (November 1964):4.

5. Interviews with participants, November 1974, February 1975, March 1975.

6. Working Group, minutes of 47th session, 3 March 1965, pp. 10–11 and appendix to item 9 of the agenda (unpublished).

7. See note 6 and Working Group, minutes of 49th session, 16 July 1965, appendix 5, pp. 1–2 (unpublished).

8. The meeting was held as a session of the ad hoc committee German Breeder-Prototype Program of the working party I: Science and Technology; see Deutsches Atomforum, *Tätigkeitsbericht Juli 1964–August 1965,* n.d., pp. 6–7.

9. German Atom Commission, appendix 2, minutes of 16th session, 20 May 1965, pp. 9–10 (unpublished).

10. P. Kilian and F. Amon, "Beteiligung der Industrie am Projekt Schneller Brüter mit Dampfkühlung," *Atomwirtschaft* 11 (1966):305–307; H. Kornbichler, "Zur Entwicklung des dampfgekühlten Schnellen Brüters," *Atomwirtschaft* 14 (1969):197–200.

11. Interview with a participant, February 1975.

12. E. Guthmann and C. Held, "Beteiligung der Industrie am Projekt Schneller Brüter mit Natriumkühlung," *Atomwirtschaft* 11 (1966):302–304.

13. Interviews with participants, January 1975, September 1976.

14. Kernforschungszentrum Karlsruhe, Projekt Schneller Brüter, "Ausführliche Erläuterungen zu den Anträgen zur Bereitstellung der Mittel für die Erstellung der Unterlagen zum Bau der Prototypen des Schnellen Brüters," October 1965 (unpublished report).

15. Working Group, minutes of 50th session, 3 December 1965, pp. 4–6; Working Group, minutes of 51st session, 4 February 1966, pp. 8–9; Expert Commission, minutes of 19th session, 7 February 1966, pp. 6–7; German Atom Commission, minutes of 17th session, 14 March 1966; Working Group, minutes of 52d session, 13 May 1966, pp. 6–7 (unpublished).

16. Interview with a participant, February 1975.

17. Interview with a participant, February 1975; see also H. Nau, *National Politics and International Technology: Nuclear Reactor Development in Western Europe* (Baltimore and London: Johns Hopkins University Press, 1974), p. 226.

18. Interview with a participant, February 1975; see also Nau, op. cit., note 17, p. 229, and C. Layton, *European Advanced Technology: A Programme for Integration* (London: Allen and Unwin, 1969), p. 117.

19. Interviews with participants, January 1975, September 1976.

20. Working Group, minutes of 47th session, 3 March 1965, p. 10; German Atom Commission, minutes of 16th session, 20th May 1965, p. 8 and appendix 2, pp. 9–10 (unpublished).

21. Interviews with participants, February 1975, November 1975, February 1976.

22. Kernforschungszentrum Karlsruhe, op. cit., note 14.

23. In the business year 1964–1965 or 1965, AEG Telefunken had a sales total (including tax) of DM4,135 million, Siemens DM7,179 million, and the GHH group DM4,241 million. The parent companies of Interatom as of 1966 were Demag (with total sales of DM1,061 million), Deutsche Babcock & Wilcox (DM1,926 million), and North American Aviation ($2.011 million or DM8,044 million). The sales figures are from *Handbuch der deutschen Aktiengesellschaften* (Darmstadt: Hoppenstedt, 1965 and subsequent years), and "The Fortune Directory of the 500 Largest U.S. Industrial Corporations," *Fortune* (July 1966):230–264.

24. H. Echterhoff-Severitt, "Wissenschaftsaufwendungen in der Bundesrepublik Deutschland. Folge 3: Aufwendungen der Wirtschaft für Forschung und Entwicklung im Jahre 1967," *Wirtschaft und Wissenschaft* 17, no. 3 (May/June 1969):19–22, here table 5, p. 22; idem, "Wissenschaftsaufwendungen in der Bundesrepublik Deutschland. Folge 5: Aufwendungen der Unternehmen für Forschung und Entwicklung im Jahre 1967," *Wirtschaft und Wissenschaft* 17, no. 5 (September/October 1969): 19–22, here table 10, p. 19.

25. Kernforschungszentrum Karlsruhe, op. cit., note 14, p. 5.

26. Interviews with participants, February 1975, February 1976.

27. Remarks by J. Pretsch in a discussion reported in *Arbeitsgemeinschaft für Forschung des Landes Nordrhein-Westfalen, Natur-, Ingenieur- und Geisteswissenschaften,* no. 163 (Cologne and Opladen: Westdeutscher Verlag, 1966), pp. 77–98.

28. Interviews with participants, December 1974, February 1975, November 1975, September 1976, October 1976.

29. U.S. Atomic Energy Commission, *Civilian Nuclear Power . . . a Report to the President, 1962,* November 1962.

30. "The Dominant Position of Westinghouse and General Electric," *Nucleonics Week* (9 May 1963):6.

31. "AEC Asks Bids on 4 Prototypes. . . . ," *Nucleonics* (April 1964): 25; "Advanced Converters. . . . ," *Nucleonics* (August 1964):53; "AEC Has Prospects for 3 Advanced Converter Prototypes," *Nucleonics* (September 1964):20.

32. "202 Industry Phase Assays AEC Converter, Breeder Reactor Planning," *Nucleonics Week* (11 April 1963):1–5.

33. See note 31.

34. "Pittman Clarifies AEC Stand on Converters, Breeders," *Nucleonics* (July 1964):27.

35. "AEC Spells Out 26-Year Fast Breeder Development Plan," *Nucleonics* (January 1964):22.

36. "AEC has Prospects for 3 Advanced Converter Prototypes," *Nucleonics* (September 1964):20.

37. "GE Developing Two New Long-Range Reactor Concepts," *Nucleonics Week* (28 June 1962):1-2; "Southwest Group Gets Six New Advanced Reactor Development Proposals," *Nucleonics Week* (16 August 1962):1; "AI and GE Compete for SAEA Advanced Reactor Development Contract," *Nucleonics Weeks* (27 September 1962):1; "Southwest Group, with Germans, to Build GE Fast Oxide Experiment," *Nucleonics Week* (28 March 1963):1-2.

38. See note 32.

39. "GE: Commercial Breeder by '74," *Nucleonics* (November 1964): 18. Reprinted with permission.

40. In addition to the above references this account is based on "General Electric," *Nuclear News* (June 1967):25-26; "General Electric: Major Fast Breeder Projects," *Nuclear News* (August 1967):35-38; "Karl P. Cohen," *Nuclear News* (July 1968):24-29.

41. Until the end of the project, these cost-overruns added up to about $6 million; see W. Häfele, "Ergebnis und Sinn des SEFOR-Experiments," in *Einheit und Vielfalt,* Festschrift für C.F. von Weizsäcker zum 60. Geburtstag, edited by E. Scheibe and G. Süssmann (Göttingen: Vandenhoek and Ruprecht, 1973), pp. 248-269; W. Schikarski, "SEFOR—ein internationales Projekt der deutschen Schnellbrüter-Entwicklung," *Atomwirtschaft* 14 (1969):414-416.

42. "The SEFOR Contract," *Nuclear News* (June 1969):29.

43. "Westinghouse: About-Face on Breeder Prototype," and "GE: 2 Prototypes for '69," *Nucleonics Week* (24 March 1966):7-8.

44. "AEC Divides Reactor Development Staff. . . . ," *Nuclear News* (26 November 1964):1; "AEC Breeder Program," *Nuclear News* (24 February 1966):6-7; "Liquid-Metal Fast Breeder: . . . ," *Nuclear News* (10 March 1966):3-4; "Wider Industry Initiative on Breeders," *Nucleonics Week* (24 March 1966):5.

45. "ESADA:GE:LMFBR," *Nuclear News* (September 1967):16.

46. "GE/ECNG One-Year Steam-Cooled Reactor Study," *Nuclear News* (April 1967):7.

47. "GE: 2 Prototypes for '69," op. cit., note 43.

48. "GE Group to Explore Nuclear Frontiers," *Nuclear News* (May 1965):5; W. Häfele, "Schnelle Brutreaktoren, ihr Prinzip, ihre Entwicklung und ihre Rolle in einer Energiewirtschaft," in *Arbeitsgemeinschaft für Forschung des Landes Nordrhein-Westfalen, Natur-, Ingenieur- und Gesell-*

schaftswissenschaften, no. 163 (Cologne and Opladen: Westdeutscher Verlag, 1966), pp. 49-74.

49. "Westinghouse," op. cit., note 43.

50. "The Next Step Is the Breeder Reactor," *Fortune* (March 1967): 121-123, 198-202. Reprinted with permission.

51. "Und Deutschlands Schnelle Brüter?" *Sonntagsblatt* (10 October 1965); "Speed-Up on Breeders Urged in Germany," *Nucleonics* (November 1965):26-27; "Die Dampfkühlung setzt sich durch," *Frankfurter Allgemeine Zeitung* (26 January 1966); "Mit 'Argonaut', Plutonium und Neutronen. . . ," *Die Welt* (29 January 1966); "Karlsruhe Fast Breeder Project," *The Engineer* (25 February 1966):331-335. See also "Germany to Build Two Breeder Types," *Nucleonics* (March 1966):28; W. Häfele, op. cit., note 48, p. 68.

52. For a lucid discussion of bias in R&D project evaluation see C. Freeman, *The Economics of Industrial Innovation* (Harmondsworth: Penguin, 1974), pp. 222-254.

53. A.W. Marshall and W.H. Meckling, "Predictability of the Costs, Time, and Success of Development," in *The Rate and Direction of Inventive Activity,* edited by National Bureau for Economic Research (Princeton, N.J.: Princeton University Press, 1962), pp. 461-475.

54. M.J. Peck and F.M. Scherer, *The Weapons Acquisition Process: An Economic Analysis* (Boston: Harvard Graduate School of Business Administration, 1962), pp. 428-460.

55. D.H. Allen and P.J. November, "A Practical Study of the Accuracy of Forecasts in Novel Projects," *The Chemical Engineer* (June 1969): 252-262.

56. H. Thomas, "Some Evidence on the Accuracy of Forecasts in R&D Projects," *R&D Management* 1 (1971):55-69.

57. Interviews with participants, December 1974, February 1975, December 1975, September 1976, October 1976.

58. Some papers by O. Löbl, a consultant to RWE, give an illustrative reading on the utilities' view on such questions with regard to the commercial introduction of first-generation nuclear power plants; see "Wirtschaftlichkeit der Atomenergie," *Elektrotechnische Zeitschrift* (1 July 1956):460-464; "Erzeugungskosten des Atomstroms," in *Kerntechnik,* edited by W. Riezler and W. Walcher (Stuttgart: Teubner, 1958), pp. 903-929; "Die Kostenfaktoren der Atomenergie," *Energiewirtschaftliche Tagesfragen* 9 (1959/60):153-157; "Streitfragen bei der Kostenberechnung des Atomstroms," in *Arbeitsgemeinschaft für Forschung des Landes Nordrhein-Westfalen,* no. 93 (Cologne and Opladen: Westdeutscher Verlag, 1961), pp. 7-19, 57-77.

59. H. Kornbichler, "Wo steht die deutsche Schnellbrüterentwicklung," *Atomwirtschaft* 16 (1971):195-208. Reprinted with permission.

60. Kernforschungszentrum Karlsruhe, op. cit., note 14, pp. 1–5.

61. H. Grümm, D. Gupta, W. Häfele, E. Schmidt, and J. Seetzen, "Demand for Nuclear Fuels and Costs of Different Reactor Types in Germany," Report KFK-366 (Karlsruhe: Gesellschaft für Kernforschung, September 1965).

62. Interviews with participants, November 1975, December 1975, October 1979, March 1980.

63. General Electric Company, "Liquid Metal Fast Breeder Reactor Design Study," Report GEAP-4418, January 1964.

64. Smidt et al., op. cit., note 3.

65. M.W. Rosenthal et al., "A Comparative Evaluation of Advanced Converters," Report ORNL-3683 (Oak Ridge, Tenn.: Oak Ridge National Laboratory, January 1965).

66. G.D. Collins, "Comparative Study of PuC-UC and PuO_2-UO_2 as Fast Reactor Fuel. Part II: Economic Considerations," Report GEAP-3880, November 1962.

67. Interviews with participants, October 1979, March 1980.

68. See note 65.

69. H. Grümm et al., "Ergänzendes Material zum Bericht Kernbrennstoffbedarf und Kosten verschiedener Reaktortypen in Deutschland' (KFK-366)," Report KFK-466 (Karlsruhe: Gesellschaft für Kernforschung, September 1966), pp. 5–8 to 5–9.

70. Kernforschungszentrum Karlsruhe, op. cit., note 14; W. Häfele, "Das Projekt Schneller Brüter Karlsruhe," *Atomwirtschaft* 11 (1966):293–302; idem, "Statusbericht 1970 des Projekts Schneller Brüter," *Atomwirtschaft* 15 (1970):171–175; E.G. Schlechtendahl, "Stand und Fortgang des Projekts Schneller Brüter," *Atomwirtschaft* 13 (1968):377–378.

71. F. Baumgärtner and W. Ochsenfeld, "Development and Status of LMFBR Fuel Reprocessing in the Federal Republic of Germany," Report KFK-2301 (Karlsruhe: Gesellschaft für Kernforschung, May 1976); "Pu-Brennstoff erstmals in KfK aufgearbeitet," *Atomwirtschaft* 24 (1979):212.

72. Interviews with participants, February 1975.

5 Termination of the Steam-Cooled Fast Breeder Reactor

Prelude

The decision in 1966 to start the design of a 300-MWe prototype plant put West German activities on the steam-cooled fast breeder far ahead of those in other countries.[1] Outside West Germany, R&D efforts on this technology were largely confined to conceptual design studies. Very little was done on component development and testing. An experimental plant of a steam-cooled fast breeder had not as yet been built anywhere. In Europe, the steam-cooled fast breeder was investigated by Belgonucléaire and by the Swedish AB Atomenergi, both firms under state control or living on government-financed R&D.[2] Some preliminary studies were also done in Britain. In the United States, Babcock & Wilcox, up to 1965, had made a small effort on this concept, initially self-financed, later with utility support.[3] The General Electric (GE) Company was prompted by talks with the Karlsruhe group to start some design studies on its own.[4]

The most advanced effort abroad in the 1960s was a one-year feasibility study for a 50-MWe reactor by GE, commissioned in the spring of 1967 by the East Central Nuclear Group, an association of fourteen American utilities.[5] It was funded with $1.2 million. This sum is about one-tenth of what the West German government had committed to prototype design, not including the R&D work performed at Karlsruhe. The American project was in a precarious situation, as the U.S. Atomic Energy Commission (AEC) showed little interest in the steam-cooled fast breeder. After a review of the fast breeder program in 1965–1966, the AEC had given a clear priority to the sodium-cooled fast reactor. The steam-cooled as well as the helium-cooled versions were conceded only a backup function.[6] GE and its utility sponsors therefore sought international collaboration for their project and in 1966 made such a proposal to the Germans. The Karlsruhe project management favored German participation with a contribution of DM15 million. The ministry, however, wanted to proceed only if AEG participated with some of its own funds. AEG declared in October 1967 that it would not commit its own money. Thereafter the proposal was abandoned without any further discussion.[7] This decision was a harbinger of the later policy reversal in the West German steam-cooled fast breeder program. It also foreshadowed the way in which this policy reversal was brought about.

121

Problems

According to the accelerated program of 1966, the steam-cooled and the sodium-cooled versions of the fast breeder were to be developed in parallel. It was generally expected in 1966 that 300-MWe prototype plants would be constructed for both types, though a firm commitment was not yet made at that time. The final decision was to be taken after the two consortia submitted their bids. These were due in December 1969.

The Karlsruhe project management was initially convinced that both types had similar prospects for success and that both would find a place on the market.[8] Because it expected commercial steam-cooled fast breeders to have lower unit sizes and lower capital costs than sodium-cooled fast breeders, it estimated that steam-cooled fast breeders would be preferable for operation in middle-load range, and liquid-metal fast breeders for base-load operation.[9] Utility speakers, however, publicly rejected this assessment, stating that breeder reactors would be deployed without exception in base-load operation.[10] This implied that there were not specific markets for each of the two fast breeder types, but that in fact these were competitors on the same market.

By the end of 1966 ministry officials had first doubts about the wisdom of leaving the final choice between the two fast breeder types to the market.[11] The precedent of the AEC's priority decision for the sodium-cooled fast breeder suggested that a choice be made at some point before market introduction, and possibly even before construction of the 300-MWe prototypes. There was also, for some officials, a growing impression that development costs might be higher than estimated and that the large funds required for a parallel full-scale development would be difficult to obtain. The ministry's draft for the nuclear program of 1968–1972 stated that possibly only one of the two fast breeder prototypes would be constructed, and perhaps none at all. Although the final version of the program used much softer language, participants in the policymaking process were alerted about these possibilities. At the Karlsruhe laboratory, the sense of competition between the proponents of the sodium-cooled fast breeder and those of the steam-cooled type grew. Results or problems in development work were seen with a growing awareness of their impact upon the competitive situation between the two fast breeder types.

Technical problems for the steam-cooled fast breeder began in October 1967, less than one year after the industrial consortia had begun their work. At an international conference held in Karlsruhe, British scientists reported on measurements correcting the previously accepted data of a certain physical constant of plutonium, the alpha value.[12] The new data reduced the breeding ratio of fast breeder reactors (FBRs) by seven to nine points in the second decimal. This would have hurt the steam-cooled fast breeder much more than its sodium-cooled competitor, as the breeding ratio for steam-cooled breeders was estimated before in Karlsruhe at 1.11 to 1.16 compared with 1.38 for the sodium-cooled breeder.[13] With a breeding ratio of 1.02 to 1.09, the steam-cooled fast breeder would therefore hardly breed at all. Real breeding seemed possible only by developing a new type of cladding for fuel rods. But this would have required an increased R&D effort.[14] The so-called alpha event lost, however, some of its importance in the course of 1968 when new measurements at Oak Ridge National Laboratory (ORNL) brought corrections in the opposite direction. The decrease of the breeding ratio then appeared only half as much as it had before.[15]

Other technical difficulties related to the behavior of nickel alloy in fast neutron flux. Nickel was thought to be a necessary ingredient in cladding material for steam-cooled breeders in order to make it resistant to the corrosive effect of high-temperature steam. Experiments by GE in the EBR-II reactor had shown, however, that such material had high failing rates in fast neutron flux. Opponents to the steam-cooled fast breeder did not doubt that these difficulties could be solved, but they thought that much additional R&D would be required to make the steam-cooled breeder feasible.[16]

More momentous problems arose at the end of 1967 with regard to the envisaged test facility for fuel elements. After preliminary tests in the VAK reactor at Kahl and the British DFR reactor, the final testing of the fuel elements for the steam-cooled fast breeder was expected from irradiations in the HDR reactor, when refurbished with a fast zone in the core. However, the fast breeder team at AEG discovered that safety-related problems prevented the planned modification of the HDR reactor. The modified reactor would have no negative Doppler coefficient. Thus it lacked inherent safety, as other measures for inherent safety were not deemed reliable enough. Further problems were caused by uncontrolled power peaks at the border between the fast and the thermal zone.[17]

The AEG team then considered the option of converting the whole core of the HDR to a fast one instead of inserting only a fast inner zone. This concept was found to provide the safety margin required, but the projected costs of about DM200 million were of the order of what construction of a new test facility would have cost. Therefore it appeared more economic to use this money for construction of a new facility. Construction of the 300-MWe steam-cooled prototype had to be deferred by the time required to design and build the smaller test facility. This meant a delay of several years in the development schedule of the steam-cooled FBR.

A wider public was made aware of the new situation by the status report on the fast breeder project on 7 March 1968 in Karlsruhe. From a discussion of the so-called alpha event and of the difficulties in providing a test bed for steam-cooled fuel elements, the project manager concluded that one should consider abandoning the parallel development of two types of fast breeders in favor of the sodium-cooled breeder.[18] This meant that the steam-cooled fast breeder was to be developed with some time lag against the sodium-cooled breeder.

In April 1968, the situation was discussed by the Fast Breeder Project Committee (see chapter 4). AEG proposed terminating design work on the 300-MWe prototype in favor of designing a small reactor for testing fuel elements. The committee agreed to this proposal. This did not imply a positive decision on the construction of the small reactor, which was made contingent upon additional information on the fuel element. For a final evaluation AEG was to make a study of a 1,000-MWe steam-cooled fast breeder, so a comparison with the potential of a 1,000-MWe boiling-water reactor (BWR) would be possible.[19]

Soon afterward, however, ministry officials began to think about more serious policy changes. Cost-overruns in other projects of the ministry, as well as increasing cost estimates in foreign fast breeder projects, had made the officials revise upward their projections of the development costs of either type. Doubts about finding a market for two different types of FBRs had increased. The steam-cooled fast breeder now lagged behind the sodium-cooled type by about six years and would find it very difficult to compete against an already commercially established breeder technology.

On top of this came bad news from the United States about an imminent termination of GE's steam-cooled fast-breeder project. For finance of the planned 50-MWe Experimental Steam-Cooled Reactor, GE and the East Central Nuclear Group had sought the support of the AEC and the collaboration of foreign countries. The utility group was prepared to cover 50 percent of the construction costs, estimated at $50 million to $100 million. When the feasibility study was completed in spring 1968, the AEC, however, refused to give support. The invited foreign countries, including West Germany, Italy, Switzerland, Sweden, and Japan, did not show much interest. Faced with lack of interest by the AEC as well as possible foreign partners, GE and the East Central Nuclear Group had to either finance the project fully out of their own funds or abandon it. The latter option was chosen. Termination of the project was announced at the end of July 1968.[20] To insiders of the fast breeder scene, lack of interest by the AEC and possible foreign collaborators had already become visible during the previous year. GE's unwillingness to put up its own funds was communicated to Karlsruhe some months before it was publicly announced.

West German ministry officials recognized that without any information exchange with foreign programs, the German project was likely to run larger risks and to incur higher costs. This realization now made them actively seek the termination of the steam-cooled fast breeder. By June 1968 the internal consultations of the ministry had proceeded to a point where the minister could announce that "a thorough review of the whole concept and of the schedule" of the steam-cooled fast breeder was to be made.[21]

In the ministry, the move to terminate the steam-cooled fast breeder did not give rise to major controversies, though personal views of civil servants differed markedly in the weight they assigned to the various arguments.[22] One key official was persuaded mainly by considerations of safety problems of the steam-cooled breeder. He was convinced that safety could never be assured reliably for a reactor in which small fuel pins with very thin cladding had to withstand the high pressures and temperatures that prevailed in a steam-cooled breeder. This argument overrode all others, though he also considered technological difficulties and market prospects as viable reasons. The international situation seemed no problem to him, and he would have seen no difficulty in obtaining from parliament the additional funds necessary for continuation of development of steam-cooled breeders. He thought that parliament would have been prepared to spend DM1,000 million on a technology which could be presented as a genuine West German development having a unique chance by being developed only in West Germany.

Another key official placed more emphasis on the financial side. In the spring of 1968 he was convinced that funds for two prototypes would not be available. With this background, the consequences of the international isolation of the West German steam-cooled fast breeder program appeared decisive.

Conflict and Consensus

When the scientists in the Karlsruhe laboratory became aware of GE's situation and of the ministry's deliberations, a bitter controversy developed between proponents of the sodium-cooled FBR and those of the steam-cooled FBR. It was exacerbated by earlier tensions between the engineering-oriented individuals working on the steam-cooled fast breeder and the physics-oriented individuals working on the sodium-cooled fast breeder.[23]

The very structure of the Karlsruhe laboratory facilitated controversy. It consisted of about a dozen rather independent institutes, each headed by a director. A foreign visitor compared it to a great cathedral, with war raging among the chapels. Some institute directors had their own preferences with regard to the choice of coolant for fast breeders. Wolf Häfele, director of the Institute of Applied Reactor Physics and manager of the fast breeder project, favored the sodium-cooled type, though he had considered the steam-cooled breeder as a real alternative as long as GE had been pursuing it. Ludolf Ritz, head of the Institute of Reactor Components, supported the steam-cooled fast breeder, which was an idea he had himself brought to Karlsruhe. Karl Wirtz, head of the Institute of Neutron Physics and Reactor Technology, had reservations about both types, favoring the gas-cooled fast breeder. However, he viewed the gas-cooled type as a back up, not an alternative to the sodium-cooled fast breeder.

On the level above the institutes, formal responsibilities were not clearly defined. The basic decisions were the responsibility of the supervisory board. Since the merger of the Kernreaktor Bau- and Betriebs-GmbH into the Gesellschaft für Kernforschung mbH in 1964 the members of the supervisory board were named by the Federal Government and the Land Government. The policy of this board was determined by the Federal Ministry for Scientific Research.

Three administrators had overall management of the center. A scientific council comprising all institute directors of the center and a few additional scientists was to advise the supervisory board, whereas its role vis-à-vis the administrators was not clearly defined. The project manager was formally under the direction and control of the scientific council. In practice, however, this control was not very tight, particularly in matters for which the project manager had the ministry's consent. With regard to access to the policymaking process, the proponents of the steam-cooled fast breeder in the Karlsruhe laboratory were at a disadvantage, since they were not well placed in the advisory bodies of the ministry. None of them was in the Working Group of the German Atom Commission or on the Fast Breeder Project Committee.[24]

The controversy over the steam-cooled FBR went into a new phase when the project manager, obviously with the ministry's consent, in July 1968 stopped further procurement orders for activities on the steam-cooled breeder, and informed AEG that the laboratory would not support any more activities on this type.[25] On 11 July the American journal *Nucleonics Week* reported (probably on the basis of information from within the Karlsruhe laboratory) that in West German nongovernment circles the steam-cooled fast breeder was considered moribund.[26] The peak of the controversy was reached when Ritz was forbidden by the laboratory's administration to present his dissenting views in a talk at a technical university, which had been arranged some time before.

The termination of the steam-cooled fast breeder was the first policy decision in the course of the fast breeder program which gave rise to public debate. It was triggered by K. Rudzinski of the *Frankfurter Allgemeine Zeitung,* one of the larger West German newspapers. In lengthy and obviously well-informed articles, Rudzinski had argued since 1965 the advantages of steam-cooling over sodium-cooling, and had launched a number of attacks on organizational and managerial faults in the ministry as well as in the Karlsruhe laboratory.[27] When the controversy about terminating the steam-cooled FBR began, his attacks against the sodium-cooled FBR and the Karlsruhe project management grew more acute. Some articles even provoked parliamentary questions.[28]

The fast breeder team at AEG fought for continuation of its work. At the project committee's next meeting on 17 September, it presented a "shortened steam-cooled fast breeder program," which envisaged a 200-MWt test reactor to be followed by a 500-MWe prototype plant. It proposed that building the test reactor at this relatively large unit size would allow leaping to a 500-MWe prototype plant in the next phase and thus compensate partly for the delay encountered with the test reactor. This proposal was in line with the project committee's prior recommendation to concentrate efforts on the design of the test reactor. In view of the good case made by the AEG team, the project committee did not alter its prior recommendations.

At that time ministry officials were talking with higher-level management at AEG's nuclear department in order to elicit the company's consent to terminate the project. They made it clear to company executives that the ministry was not prepared to cover increased costs of an isolated national program. AEG was thus confronted with the alternative to show some of its own funds or abandon the project. The company chose the latter. It now became aware that it was not going to make money on its light-water reactor (LWR) contracts. The top management of the company was not enthusiastic about nuclear power anyway. Under such circumstances there was little inclination to depart from the company's policy of putting no funds of its own into the steam-cooled fast breeder.[29] On 28 November 1968 an information service reported that AEG did not want to continue the steam-cooled fast breeder.[30]

At the Karlsruhe laboratory, there was no need for long discussions as the ministry had the say in the supervisory board. The final decision to terminate activities on the steam-cooled fast breeder was taken on 8 October 1968.[31]

After AEG had agreed to the termination of the steam-cooled FBR, the formal decision-making process on the industrial part of the steam-cooled fast breeder program could be initiated. As the controversy had spread into the public arena, the ministry made considerable efforts to procure political support for its decision. The Bundestag Committee for Scientific Research was informed by the minister on 24 October 1968 in his speech introducing the 1969 budget. He said that, according to the prevalent opinion, the steam-cooled fast breeder should be abandoned, and a final decision was to be taken at the end of the year.[32] Statements about an imminent revision of the steam-cooled fast breeder project were given to the press by the ministry on 6 November and 10 December 1968.[33] On 23 and 24 January 1969, a hearing was organized jointly by the ministry and the Bundestag committee.

There proponents and opponents of the steam-cooled fast breeder in the Karlsruhe laboratory as well as representatives from firms working on the technology presented their views.[34] In formal terms, such a joint hearing by a ministry and a Bundestag committee is not provided for in the West German government machinery. It is not only somewhat unusual but does not conform to separation of legislative and executive powers as laid down in the West German constitution. The ministry often referred to this constitutional principle in order to shrug off requests of members of parliament to participate in the discussions of ministerial advisory bodies. But in this specific case a joint hearing with the Budestag committee was seen as an appropriate means of procuring political support by involving this committee in the decision-making process, or giving it at least the feeling that is was involved.

Also within its advisory bodies the ministry carefully maximized political support. On 13 December 1968 the matter was discussed by the Working Group of the Atom Commission.[35] Before the meeting the ministry circulated a letter to the committee members, in which it delineated the reasons why it considered terminating the steam-cooled fast breeder. Members of the Working Group were asked to submit their opinions in writing. Their statements were circulated before the meeting. At the meeting, each member was asked to comment on the circulated statements.

Such a procedure had never been instigated before in the Working Group. Although the etiquette of such advisory bodies requires dissenting members not to criticize in public the majority's recommendations, the ministry made sure through this procedure that the recommendation it sought from the Working Group was supported by all its members, including those unable to attend this specific meeting. The Working Group finally resolved to recommend that work on the steam-cooled fast breeder should be reduced in favor of the sodium-cooled fast breeder. Current activities should be brought to a reasonable end, and only work on the fuel element should be pursued further. Furthermore, the Working Group recommended that the consequences of such a decision for the firms participating in the steam-cooled fast breeder project should be reviewed.

The last committee to be involved in the ministry's consensus building effort was the Fast Breeder Project Committee. It gave its consent on its eighth session on 28 January 1969. On 5 February the minister formalized his decision by signing a statement, which was released to the press two days later.[36] The minister decided that the development of the steam-cooled fast breeder would be abandoned, with the exception of some work on the fuel pins at the Karlsruhe laboratory. Industry was instructed to stop design work, to bring experiments and tests to a meaningful end, and to present the results in a final report.

To alleviate the impact of the termination of the steam-cooled fast breeder project on AEG, the ministry in 1969 gave three R&D contracts to the company: one for design studies on a carbide fueled sodium-cooled fast breeder reactor, one for development work on a carbide-fuel element for sodium-cooled breeders, and one for design studies for a large fast test reactor of several hundred thermal megawatts. These contracts involved the allocation of about DM7 million of government funds, which was roughly the amount saved through termination of the steam-cooled fast breeder.[37]

When these three studies were completed by 1971, AEG abandoned its breeder activities. At this time AEG and Siemens had formed Kraftwerk Union and had agreed to merge their nuclear power activities as soon as the license agreements with GE and Westinghouse allowed.

At the Karlsruhe laboratory, work on the steam-cooled FBR was reduced to activities on the fuel element. At the status report of 1971, the situation of the steam-cooled fast breeder was reviewed by speakers of the project management as well as by a panel comprising representatives of the Karlsruhe laboratory, the reactor manufacturers, a utility, and a member of the Bundestag Committee for Education and Science.[38] Considerable progress was reported which made the steam-cooled fast breeder appear in a much better light. A special pressure compensation system had been developed, which reduced the strains on the fuel clad resulting from the interior pressure of the fission gases. Investigations on austenitic steels showed that an appropriate cladding material could be found without recourse to nickel alloys.[39]

Policy was not changed, however, since the sodium-cooled fast breeder now was seen to require much more resources and its commercial use to be much more remote than anticipated two years earlier. When the Working Group in January 1971 approved a small program for work on the gas-cooled FBR funded with DM20 million ($6.5 million) over four years, the representative of AEG proposed that in parallel to a 1,000-MWe reference design of the gas-cooled FBR a similar study should be also done on the steam-cooled FBR. Pointing out that the termination of industrial work on steam-cooled fast breeders had been decided in 1969 on economic rather than technical grounds, he suggested that the steam-cooled FBR should be taken into consideration if the government was to make funds available for an alternative to the liquid-metal FBR. The Working Group recommended that the steam-cooled FBR should be based on the same footing as the gas-cooled FBR. This meant that no direct subsidies should be given to industry, but government laboratories should do supportive R&D, provided industry would do the design work on its own funds. Although AEG later submitted a detailed proposal for a DM1.2 million design study, it did not commit the company's own funds. Therefore, activities on the steam-cooled

FBR at Karlsruhe remained at a low level. When Ritz retired in 1974, they were phased out with the consent of Kraftwerk Union.[40]

Perceptions of Uncertainty

The termination of the steam-cooled FBR was the first decision in the West German program where technical matters were discussed in some detail. The statements of the various actors at the Bundestag hearing in 1969 and in the advisory bodies show a great difference in their sense of the uncertainty involved in making a technical and economic assessment. These different perceptions of uncertainty reflect different backgrounds of experience.

The Karlsruhe Majority

The proponents of the sodium-cooled FBR at Karlsruhe showed in their statements a high level of confidence in their grasp of the technical and economic prospects of the different fast breeder types.[41] In discussing the problems of the steam-cooled fast breeder, they went into great technical detail: the consequences of the new alpha value; difficulties of nickel alloys; and technical possibilities for compensating for the reduction in breeding that resulted from the new alpha values. At the same time they were quite confident in advanced developments of the sodium-cooled fast breeder, for example carbide fuel.

As to future development costs they had precise opinions, estimating the cost of precommercial plants for the steam-cooled fast breeder at DM630 million compared to DM400 million to DM500 million for the sodium-cooled fast breeder. However, in discussing the implications of an isolated national development, they regarded uncertainties in development cost to be on the order of magnitude.

Regarding market introduction, they expected a time lag of at least six years for the steam-cooled fast breeder. In discussing the electricity costs of different fast breeder types they made some concessions to uncertainty, giving ranges of probable values rather than hard figures (see figure 5-1). Nevertheless they held their estimates to be valid arguments. In proposing them, they paid much attention to short-term changes in the economics of nuclear power. For instance, they estimated that a recent reduction of American enrichment prices, together with a further trend in this direction, had reduced the difference in electricity costs between the liquid-metal fast breeder and the LWR by 0.1 DPf/kWh relative to earlier assumptions. Finally they saw a strong argument in computations showing that the steam-cooled fast breeder would conquer a much smaller market share because of its lower breeding potential.

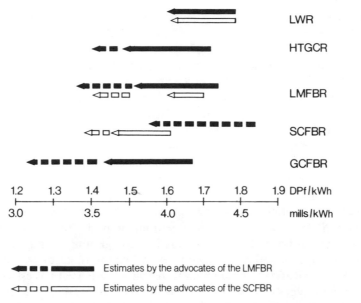

Sources: L. Ritz, "Zur Entwicklung des dampfgekühlten Schnellen Brüters," *Atomwirtschaft* 14 (1969):200–202; M. Fischer et al., "Untersuchungen der technischen und ökonomischen Situation dampfgekühlter schneller Brutreaktoren," Report KFK-918 (Karlsruhe: Gesellschaft für Kernforschung, January 1969).

Note: Broken lines indicate electricity costs of the reactor types as employing advanced technical features that require considerable additional R&D: HTGCR with direct cycle, LMFBR with carbide fuel, SCFBR with heavy steam, GCFBR with direct cycle and carbide fuel. LMFBR advocates regarded the SCFBR with normal steam as an advanced type of technology requiring considerable additional R&D.

Figure 5–1. Electricity Costs as Estimated by Karlsruhe in 1969

Proponents of Steam Cooling

By contrast with the physics background of the Karlsruhe majority advocating the sodium-cooled fast breeder, the views of Ludolf Ritz reflect his training and experience as an engineer.[42] His arguments for the steam-cooled fast breeder were more general in character and qualitative rather than quantitative. He pointed to the conservative attitude of the manufacturers of heavy-plant equipment, who were skeptical about revolutionary innovations. In his view, this attitude was justified by the experience of failing innovations, even incremental ones. This experience was to have much weight, since the equipment involved was of large scale and operation breakdown was very costly. Steam-power plants had incurred no revolutionary technological change over a long period, during which many technical alternatives to electricity generation by steam turbines had failed.

When nuclear power plants were developed, out of many different options the LWR, as the technology closest to conventional steam technology, was successful.

Steam as the coolant option closest to LWRs should therefore also be thoroughly considered for FBRs. Ritz pointed out that in line with such considerations his work on steam-cooled breeders had been guided toward as simple a reactor as possible, using conventional temperatures and steam pressures and consisting of few components. The objective was high plant availability and low plant costs, as these factors were in his view decisive in the success of LWRs. All key components such as the evaporator and the ventilator had been constructed in his institute at small scale and tests of components and a small pilot circuit had provided good results.

As far as electricity costs were concerned, Ritz remarked that in general estimates of electricity costs were a questionable basis for an evaluation of different reactor types. He stressed that, in the beginnings of nuclear power, any reactor type looked like being the best, when the first design was put on paper. More specifically he pointed to recent experience in West Germany with different types of reactors, such as BWRs with fossil or nuclear super-heating, and the sodium-cooled thermal reactor. At first, these all promised substantial cost advantages and were later abandoned. Ritz remained confident that the steam-cooled fast breeder would undercut the costs of electricity of oxide-fueled sodium-cooled fast breeders. As a concession to the terms of debate set by his opponents, he even gave some quantitative estimates (see figure 5-1).

Industry

Kornbichler, the speaker of AEG, regarded the termination of the steam-cooled fast breeder abroad as "the only really decisive change" since the inception of work on this reactor in West Germany.[43] He pointed out that the isolated development of a reactor increased technological risks, as experience with the new technology would accrue much more slowly, and in the case of a serious technological setback it was less probable that a solution would be found. Kornbichler stressed that no type of reactor so far had been developed in West Germany independently from foreign know-how.

As to development costs, Kornbichler estimated work till completion of the prototype would require about DM1,000 million, this estimate being on the optimistic side. In addition he noted that there existed technological uncertainties which, if they turned out badly, would push costs very much higher and would necessitate a development effort beyond present capacity. This would happen, for instance, if the cladding materials envisaged so far for steam-cooled breeder fuel pins proved inappropriate or if the contamination of the cooling circuit could not be held at a level allowing a direct cooling cycle.

An important argument related to safety. A steam-cooled fast breeder developed by West Germany in isolation would negate the possibility of drawing on foreign know-how and experience with regard to engineered safeguards. The licensing procedure would be much more difficult, as the licensing authorities could not draw on operating experience abroad.

As to market prospects, Kornbichler showed marked skepticism in electricity-cost estimates that are made at an early stage of the development process. He pointed out great uncertainties in the commercialization of a steam-cooled breeder being developed independently in West Germany. Commercial introduction of LWRs in West Germany was very slow, though a good number of such reactors were already operating in the world. It appeared therefore presumptuous to bring such an independently developed reactor on the market. He suggested that even the sodium-cooled breeder, assuming successful development, would have difficulty getting into the market. It must be remembered that this was not the original position of AEG as proposed to the project committee in September 1968, but the outcome of later discussions between the ministry and his company.

Utilities

Since there were no presentations by utility spokesmen at the hearings on 23 and 24 January 1969, the utilities' view may be grasped from the statements of their representatives in the Working Group of the German Atom Commission.[44] With regard to the estimates of electricity costs in various types of FBRs as well as other unproven types of nuclear plants, they noted that the scope of uncertainties in these estimates was so wide that a choice between different types of plant was not possible on this basis. Even for an advanced type of sodium-cooled FBR using carbide fuel, they did not see a certain market advantage. Concerning the different coolant options for FBRs, they deemed the state of knowledge to be such that no final judgment about the best solution could be given at that time. There was a possibility that the sodium-cooled and helium-cooled versions would someday run into difficulties which might necessitate reevaluation of the steam-cooled fast breeder.

The apparent time lag of the steam-cooled type against the sodium-cooled type was not regarded as decisive by utility representatives. They deemed it quite possible that the sodium-cooled type might also suffer substantial delays before commercial application. In a longer time horizon, the increased development cost of the steam-cooled fast breeder did not present an overriding argument, as the costs might be much less spectacular if the steam-cooled fast breeder was not developed at once, but as an evolution from the nuclear superheat BWR.

The long doubling time of the steam-cooled fast breeder was regarded as a real disadvantage versus the sodium-cooled fast breeder. Utility representatives saw the main argument in favor of a termination of the steam-cooled fast breeder in the international situation.

The Ministry

The justifications given in the statement of the Minister for Scientific Research coincide roughly with the arguments given by the opponents to the steam-cooled fast breeder at Karlsruhe and by supplier industry.[45] This could be expected, as an institution may justify a decision by any available good reason.

A closer analysis, however, shows some points where the ministry shared the more clear-cut views of the Karlsruhe majority rather than the more cautious views of supplier industry or utilities. As to the six-year time lag of the steam-cooled breeder, the ministry thought it could be reduced if sufficient effort were made on R&D. But it stressed that even a small delay would mean the steam-cooled breeder would have to compete with an established sodium-cooled breeder. Additional government money would then have to be spent on a demonstration plant in order to make this type commercially viable.

As to estimates of electricity costs, the ministry adopted the position that there would be no differences in the electricity costs between the steam-cooled, the sodium-cooled, and the helium-cooled fast breeders. The ministry mentioned the disadvantage of the steam-cooled breeder in having a low breeding potential, but a more critical fault in the steam-cooled breeder was its lack of potential for further technical and economic advance comparable to carbide fuel in sodium-cooled and helium-cooled breeders.

Within the spectrum of differing sensitivities to uncertainty, the ministry chose to settle in the middle. It showed a higher appreciation for uncertainty than the Karlsruhe majority, but did not accept the broad scope of uncertainty as suggested by reactor manufacturers and utilities. Such a middle position commended itself for the ministry's effort to achieve as large a consensus as possible. But it was more than tactical. It was a sign that, contrary to the 1966 decision, the ministry no longer uncritically accepted the views proposed by the Karlsruhe scientists. Of course, the clash of opinions at Karlsruhe contributed to the ministry's critical attitude. More importantly, ministry officials now listened carefully to the less optimistic and more cautious opinions of utility representatives. This enabled them to finally arrive at an independent judgment, which permitted them to use financial levers in order to clinch AEG's consent.

Hindsight

Further experience in the West German fast breeder program fully vindicated the caution and skepticism of the reactor manufacturers and utilities. The economic and technical information coming to light in 1970 to 1972 during the negotiations on the construction contracts for the 300-MWe liquid-metal prototype SNR-300 added drastically to the estimated costs of this plant and of the whole program, dealing a blow to earlier optimistic forecasts of electricity costs. This confirmed the utilities' and reactor manufacturers' reservations on cost estimates based on paperwork. Such reservations also proved valid with regard to models of future electricity economies based on electricity-cost estimates. Given the uncertainties involved, the Karlsruhe majority seems to have overstated the value of these models when using them as an argument for terminating the steam-cooled fast breeder. Nevertheless, such models appear helpful in exploring technical as opposed to cost factors affecting the future market of specific reactor types, for example, the availability of plutonium.

As for the time scale of fast breeder commercialization, a six-year delay of the steam-cooled fast breeder against the sodium-cooled fast breeder proved to be a spurious argument. In the context of the reassessment of fast breeder economics as necessitated by the SNR-300, the schedule for the commercial introduction of fast breeders was delayed by one or two decades. The estimated DM130 million to DM180 million difference in the total development cost of the steam-cooled versus the sodium-cooled fast breeder proved negligible in the light of later cost increases.

The international isolation of the West German steam-cooled fast breeder project, which for the manufacturing and utility industries was the decisive argument, has gained weight in later years. As the development of the sodium-cooled type went on, it appeared that R&D tasks were much bigger than anticipated, particularly in safety-related issues, such as the hypothetical core-disruptive accident. A considerable amount of international cooperation has evolved in this field. Although some of this research relating to the sodium-cooled type may be applicable also to the steam-cooled type, the difficult licensing procedure for the SNR-300 and the amount of additional R&D it required, have validated the apprehensions about an isolated national development. In view of this situation, even scientists who formerly advocated the steam-cooled fast breeder now regard its termination as irreversible.[46]

Costs of an Accelerated Schedule

The international situation of the steam-cooled fast breeder in 1969 was not significantly different from that in 1966: no prototype of this reactor type was in operation or under construction anywhere in the world, and design work was of a preliminary kind. The most advanced project abroad, the $1.2 million feasibility study on a 50-MWe reactor by GE for the East Central Nuclear Group, was begun after the West Germans had embarked on the design of a 300-MWe prototype. Thus the main argument for terminating the steam-cooled fast breeder already existed in 1966. In this regard, the 1969 decision to abandon this technology was not the result of new information but the better judgment of the decision makers, particularly the ministry.

With regard to the other technical arguments used in the 1969 decision, one can observe that they were based on information produced abroad, not under the German program. This information would have been available without the German commitment to prototype design. One can conclude that if the West German activities had proceeded in a sequential manner, that is, if design of the 300-MWe prototype had been deferred until the design and construction of the small test facility was completed, the German program would have had the same amount of information available in 1969, with an expenditure of about DM30 million less.

Notes

1. For a short survey of activities outside West Germany see B. Affeldt et al., "Entwicklungsprogramm: Dampfgekühlter Schneller Reaktor. Zusammenfassender Bericht. 1. Teil. Technische Beschreibung," Report BMBW-FB K 70-20 (Bonn: Bundesministerium für Bildung und Wissenschaft, October 1970), p. 2-2.

2. G. Tavernier et al., "HERMES, Réacteur Surchauffeur à Neutrons Rapide," Paper P/516, *Third United Nations Conference on the Peaceful Uses of Atomic Energy, Geneva 1964,* proceedings, vol. 6, pp. 62–69; G. Tavernier et al., "HERMES, a Modular Steam Cooled Fast Breeder Concept," proceedings of the National Topical Meeting on Fast Reactor Technology, 26–28 April 1965, Report ANS-100, American Nuclear Society, pp. 99–104. For Swedish activities see remarks of P.H. Margen and K. Hannerz, in *3rd Foratom Congress, London, 24–26 April 1967; Industrial Aspects of a Fast Breeder Reactor Programme, Discussion* (London: British Nuclear Forum, 1967), pp. 5–6, 24.

3. P.F. Schutt, "Babcock & Wilcox: Breeder Development," *Nuclear*

News (August 1967):28–30; "Steam Breeder: B & W, ECNG in New Study," *Nucleonics* (September 1964):26.

4. Interview with a participant, November 1975; see also J. Pretsch, discussion, in *Arbeitsgemeinschaft für Forschung des Landes Nordrhein-Westfalen, Natur-, Ingenieur- und Gesellschaftswissenschaften,* no. 163 (Cologne and Opladen: Westdeutscher Verlag, 1966), p. 87.

5. "GE/ECNG One-Year Steam-Cooled Reactor Study," *Nuclear News* (April 1967):7; "General Electric: Major Fast Breeder Projects," *Nuclear News* (August 1967):35–38.

6. M. Shaw, "Fast Breeder Reactor Programme in the United States," Paper 1/4, *British Nuclear Energy Society London Conference on Fast Breeder Reactors, 17–19 May 1966,* proceedings, pp. 36–65.

7. Working Group III/1, on Reactors, of the German Atom Commission (henceforth Working Group), minutes of 52d session, 13 May 1966, p. 5; minutes of 57th session, 18 October 1967, pp. 6–7 (unpublished).

8. Working Group, minutes of 47th session, 3 March 1965, pp. 10–11, and appendix to item 9 of the agenda (unpublished).

9. W. Häfele and P. Engelmann, "Das Projekt Schneller Brüter," in *10 Jahre Kernforschungszentrum Karlsruhe,* edited by Gesellschaft für Kernforschung (Karlsruhe: Gesellschaft für Kernforschung, 1966), pp. 17–27.

10. Statement of A. Weckesser on behalf of H. Mandel, *3rd Foratom Congress, London, 24–26 April 1967; Industrial Aspects of a Fast Reactor Programme* (London: British Nuclear Forum, 1967), p. 52.

11. Interviews with participants, February 1975, February 1976; ad hoc committee Preparation of the Third Nuclear Program of Working Group III/1, minutes, 14 November 1966 (unpublished).

12. M.G. Schomberg, M.G. Sowerby, and F.W. Evans, "A New Method of Measuring Alpha (E) for Pu 239," *IAEA Symposium on Fast Reactor Physics, Karlsruhe 1967,* proceedings, vol. 1, p. 289, quoted by W. Häfele et al., "Fast Breeder Reactors," *Annual Review of Nuclear Science* 20 (1970):393–434.

13. D. Smidt et al., "Referenzstudie für den 1 000 MWe natrium-gekühlten Schnellen Brutreaktor (Na 1)," Report KFK-299 (Karlsruhe: Gesellschaft für Kernforschung, December 1964), p. 10–6; A. Müller et al., "Referenzstudie für den dampfgekühlten Schnellen Brutreaktor (D 1)," Report KFK-392 (Karlsruhe: Gesellschaft für Kernforschung, August 1966), p. 11–2.

14. W. Häfele, "Zur Entwicklung des dampfgekühlten Schnellen Brüters," *Atomwirtschaft* 14 (1969):190–197.

15. R. Gwin et al., "Measurements of the Neutron Fission and Absorption Cross Sections of Pu-239 over the Energy Region 0.02 eV to 30

keV," Report ORNL-TM-2598 (Oak Ridge, Tenn.: Oak Ridge National Laboratory, 1969); see Häfele et al., op. cit., note 12.

16. Häfele, op. cit., note 14, p. 193; see also H. Böhm, "Zu den Brennelementproblemen beim dampfgekühlten Schnellen Brüter," *Atomwirtschaft* 14 (1969):210–211.

17. Interview with a participant, November 1975; see also H. Kornbichler, "Zur Entwicklung des dampfgekühlten Schnellen Brüters," *Atomwirtschaft* 14 (1969):197–200; A.C. Stumpe, "Entwicklungsprogramm: Schneller Kern im HDR. Zusammenfassender Bericht," Report BMBW-FB K 69–46 (Bonn: Bundesministerium für Bildung und Wissenschaft, December 1969).

18. Häfele, op. cit., note 14, p. 193; see also the editorial in *Deutscher Forschungsdienst, Sonderbericht Kernenergie* (14 March 1968):51–55.

19. Kornbichler, op. cit., note 17, p. 198; Häfele, op. cit., note 14, p. 193; interview with a participant, February 1975.

20. "Bulletin," *Nucleonics Week* (20 June 1968):3; "The Fast-Breeder Programs of Several Nations Are in the News," *Nucleonics Week* (11 July 1968):7–8; "GE and East Central Nuclear Group Have Terminated Their Steam-Breeder Effort," *Nucleonics Week* (1 August 1968):5–6. For the perception of a West German ministry official see G. Schuster, *Die reaktortechnische Entwicklung in der Bundesrepublik Deutschland,* Berichte der Studiengesellschaft zur Förderung der Kernenergieverwertung in Schiffbau und Schiffahrt e.V., no. 16 (Munich: Karl Thiemig, 1968), p. 14.

21. "Schnelle Brutreaktoren," *Pressedienst des Bundesministeriums für wissenschaftliche Forschung,* no. 11 (19 June 1968):93–94.

22. Interviews with participants, February 1975, February 1976, October 1976, May 1977, June 1977.

23. Interviews with participants, March 1975, November 1975, December 1975, October 1977, March 1980.

24. Bundesrechnungshof, "Bemerkungen zu den Bundeshaushaltsrechnungen (einschliesslich der Bundesvermögensrechnungen) für die Haushaltsjahre 1968 und 1969," Bundestagsdrucksache VI/2697, 16 September 1971, p. 98.

25. Interview with a participant, December 1975; see also "Ein FAZ-Journalist kämpft gegen das Establishment," *Capital* (February 1969).

26. "The Fast-Breeder Programs of Several Nations Are in the News," *Nucleonics Week* (11 July 1968):7–8.

27. See, for instance, "Verspielt Karlsruhe eine Chance der Atomtechnik?" *Frankfurter Allgemeine Zeitung* (FAZ) (1 June 1965); "Die Dampfkühlung setzt sich durch," *FAZ* (26 January 1966); "Der 'Natrium-Brüter'—eine Milliarden-Fehlinvestition," *FAZ* (20 July 1966); "Milliarden für den Schornstein?" *FAZ* (28 October 1968); "Die schwierigen Brutreaktoren," *FAZ* (4 December 1968); "Dampfbrüter-Brennelemente—

kein Risiko," *FAZ* (18 December 1968); "Brutreaktoren zwischen Realität und Utopie," *FAZ* (15 January 1969); "Reaktortheologie und Reaktor-realität," *FAZ* (27 January 1969); " 'Schneller Brüter' und Schwarzer Peter," *FAZ* (8 February 1969).

28. Verhandlungen des Deutschen Bundestages, 5. Wahlperiode, 196th session on 15 November 1968, Stenographische Berichte Band 68, pp. 10543–10546; see also "Bundesminister Dr. Stoltenberg in der Fragestunde des Bundestages am 8. Mai 1969," *Pressedienst des Bundesministeriums für wissenschaftliche Forschung,* no. 10/69 (14 May 1969):81–83.

29. Interviews with participants, February 1975, November 1975.

30. Editorial, *Deutscher Forschungsdienst, Sonderbericht Kernenergie* (28 November 1968):236–240.

31. Interview with a participant, December 1975.

32. Deutscher Bundestag, Ausschuss für wissenschaftliche Forschung, minutes, 24 October 1968 (unpublished).

33. "Zum Stand der Reaktorentwicklung in der Bundesrepublick," *Pressedienst des Bundesministeriums für wissenschaftliche Forschung* no. 21 (6 November 1968):186; "Die reaktortechnische Entwicklung in der Bundesrepublik Deutschland," *Pressedienst des Bundesministeriums für wissenschaftliche Forschung,* no. 24 (10 December 1968):212–217.

34. *Anhörung zum Problem der zukünftigen Arbeiten auf dem Gebiet der Schnellbrüterentwicklung in der Bundesrepublik Deutschland, 23. und 24. Januar 1969, Kurzprotokoll,* Az.: III B 1-5532-112-9/69. Part of the presentations to this hearing were later published: "Die Kontroverse über die künftige deutsche Schnellbrüter-Entwicklung," *Atomwirtschaft* 14 (1969):187–220.

35. Working Group, minutes of 60th session, 13 December 1968, pp. 3–8 and appendixes 1–12 (unpublished).

36. "Zum Reaktorprogramm der Bundesrepublik Deutschland," *Pressedienst des Bundesministeriums für wissenschaftliche Forschung,* no. 4/69 (19 February 1969):25–29.

37. Working Group, minutes of 61st session, 7 March 1969, pp. 7–10; interview with a participant, February 1975; Working Group, minutes of 61st session, 7 March 1979, p. 8 (unpublished).

38. "Wo steht die deutsche Schnellbrüterentwicklung?" *Atomwirtschaft* 16 (1971):195–208, 247–259.

39. P. Engelmann, "Die Arbeiten des Karlsruher Projektes Schneller Brüter," *Atomwirtschaft* 16 (1971):251–255. For more details see P. Engelmann and L. Ritz, "Bericht über die im KFZ Karlsruhe durchgeführten F. und E. Arbeiten zum dampfgekühlten Schnellen Brüter," Report KFK-1370 (Karlsruhe: Gesellschaft für Kernforschung, 1971).

40. Working Group, minutes of 66th session, 27 January 1971, p. 8; minutes of 67th session, 28 October 1971, p. 10 (unpublished). Interviews

with participants, December 1974. For the activities on the GCFBR see M. Dalle Donne, "The Gas-Breeder Memorandum and the Research and Development Programme of the Federal Republic of Germany in the Field of Gas-Cooled Fast Reactors," in *Gas-Cooled Fast Reactors, Proceedings of the IAEA Study Group Meeting, Minsk, 24–28 July 1972*, Report IAEA-154 (Vienna: International Atomic Energy Agency, 1973), pp. 267–288.

41. Häfele, op. cit., note 14; M. Fischer, "Zu den Brennelementproblemen beim dampfgekühlten Schnellen Brüter," *Atomwirtschaft* 14 (1969):203–206; H. Krämer and J. Seetzen, "Mögliche Entwicklungen einer künftigen Kernenergiewirtschaft in der Bundesrepublik Deutschland," *Atomwirtschaft* 14 (1969):218–220. The statement by Krämer and Seetzen summarizes the results of a study commissioned by the Federal Ministry of Scientific Research: H. Krämer and J. Seetzen, "Mögliche Entwicklungen einer künftigen Kernenergiewirtschaft in der BRD," Report JUEL-600-RG, KFK-933, Kernforschungsanlage Jülich and Kernforschungszentrum Karlsruhe, June 1969. Another study collecting arguments against the steam-cooled fast breeder was initiated from within the Karlsruhe laboratory: M. Fischer, J. Seetzen, P. Jansen, D. Faude, "Untersuchungen der technischen und ökonomischen Situation dampfgekühlter schneller Brutreaktoren," Report KFK-918 (Karlsruhe: Gesellschaft für Kernforschung, January 1969).

42. L. Ritz, "Zur Entwicklung des dampfgekühlten Schnellen Brüters," *Atomwirtschaft* 14 (1969):200–202; idem, "Zu den Brennelementproblemen beim dampfgekühlten Schnellen Brüter," *Atomwirtschaft* 14 (1969):207–210.

43. Kornbichler, op. cit., note 17.

44. Working Group, minutes of 60th session, 13 December 1968, pp. 3–8, and appendixes 4 and 7 (unpublished).

45. See note 36.

46. Interview with a participant, November 1975.

6 Construction Start of the SNR-300 Plant

The liquid-metal fast breeder fared better than its steam-cooled alternative. Due to the collaboration between West Germany, Belgium, and the Netherlands it commanded a broader technical, financial, and political basis. Moreover, it enjoyed the moral support resulting from the existence of similar programs in nearly all industrial countries.

The decision-making process relating to the construction of the SNR-300 plant was more complicated than the processes described in the earlier sections of this study. As a collaborative project, the SNR-300 involved actors from three countries, though the West Germans, paying 70 percent of the bill, were to call the tune. The SNR-300 plant underwent the normal licensing procedure for commercial nuclear power plants, so the licensing authorities were involved. Utilities as the future owner and operator of the prototype plant played a significant role for the first time in the program. Last but not least, the government funds to be committed were larger by an order of magnitude.

Because of this complexity, this chapter does not proceed in chronological order, but breaks down the decision making according to the main issues. The formal chain of events can be told in a few words. At the end of 1969, the industrial consortium submitted its design. The start of construction was delayed, however, as extensive alterations were demanded by the licensing authorities and by the utility company that was to operate the plant. The revised safety report incorporating the basic features of these alterations was submitted in June 1971. The advisory bodies of the West German Science Ministry reviewed the technical and economic issues of the project between the fall of 1970 and the end of 1971.[1] The Fast Breeder Project Committee stated on 25 October 1971 that the technology was ready for construction of a prototype, and that the government should go ahead with it, provided the licensing authorities gave their consent to a technically feasible safety concept. The Working Group III/1 on Reactors of the German Atom Commission (henceforth Working Group) recommended on 28 October 1971 that the prototype be constructed as soon as possible. On 15 December 1971, the Reactor Safety Commission resolved that the new safety concept was technically feasible. A joint government commission, comprising representatives of West Germany, Belgium, and the Netherlands, met on 7 March 1972 and decided that the prototype should be built.

After the three governments had appropriated the initial funds, the owner/operator utility issued on 23 March 1972 a letter of intent to the supplier consortium and a civil engineering consortium, and placed first part-orders amounting to DM24.4 million. On 21–22 June 1972 the Reactor Safety Commission approved the new safety concept in general. The supply contracts were signed on 10 November 1972. Their coming into force was made dependent on a number of conditions: that the governments give their consent to the contracts and formally commit themselves to provide the money through appropriation statements; that the licensing authorities grant the first part-license; and that the parent company of Interatom, Kraftwerk Union, give a guarantee for Interatom's contractual obligations. The first part-license was issued on 22 December 1972. After the other conditions were fulfilled, ground was broken on 24 April 1973 on a site on the left bank of the Rhine near Kalkar, not far from the West German-Dutch border.[2]

Organizational Setup

In line with the collaboration established in 1965–1966 on the level of government laboratories, the governments of West Germany, Belgium, and the Netherlands exchanged in 1967 memoranda that envisaged joining forces in the construction of the sodium-cooled prototype and in subsequent activities, the aim being the establishment of a joint fast breeder industry. The guidelines for the memoranda's implementation, which were negotiated a year later, laid down that West Germany would cover 75 percent of the costs of the prototype plant, Belgium and the Netherlands each 12.5 percent. They also delineated the shares of the countries in the supplies for the reactor and defined the R&D tasks to be performed in each country.[3]

On the basis of the government agreements, formal ties were established at the level of industry. On 11 January 1968, Siemens and Interatom (West Germany), Belgonucléaire (Belgium), and Neratoom (Netherlands) signed contracts for collaboration and formed a consortium (later called Konsortium SNR) for the development and construction of the planned sodium-cooled fast breeder prototype. The contracts included license agreements by which know-how was exchanged. In accordance with the previously established lines of collaboration among the government laboratories, responsibility for fuel development was shared by Siemens and Belgonucléaire, and responsibility for the cooling components, such as pumps, steam generators, and heat exchangers, was shared by Interatom and Neratoom.[4]

The members of the industrial consortium were quite different in character. Neratoom is an engineering consulting company. Its shares are held by several Dutch companies, mainly in the shipbuilding and heavy-plant equipment industries.[5] Development as well as manufacture of the sodium components for the fast breeder are done in the parent companies of Neratoom, while Neratoom's function is to provide liaison among these firms and the German and Belgian partners in the consortium. Belgonucléaire, in addition to engineering, manufactures plutonium fuel. This was a privately owned company, its main shareholder being the large mining concern Union Minière. However, the company derived its main income in the second half of the 1960s from R&D financed by the Belgian government. In 1971, the Belgian government consolidated the firm by taking over 50 percent of its shares.[6]

Interatom is an engineering company with a focus on development work involving design and testing. The company owns small production facilities for special components, mainly for uranium enrichment. Manufacture of reactor components is contracted out to subsuppliers. Since 1969, when Siemens acquired a majority holding and transferred its fast breeder activities to this company, Interatom has been the only partner on the West German side. In 1973 the firm became a fully owned subsidiary of Kraftwerk Union.[7]

Luxatom of Luxemburg joined the consortium later. It took over work on a fuel-element inspection cell, which it shared with Interatom and Belgonucléaire. Among the partner countries it was understood that the Luxemburg government would cover about 1 percent of the costs of the prototype, and industry of this country would be given a corresponding share of the supplies.[8]

From January 1968 onward, representatives of the Belgian and Dutch organizations participated regularly in the sessions of the Fast Breeder Project Committee, when dealing with the part of the project concerning the sodium-cooled FBR. Matters relating to the steam-cooled FBR were discussed in separate sessions, in which the foreign partners did not participate, though they were informed about major issues.

For construction and operation of the SNR-300 plant, an organizational setup was chosen similar to that for the three West German light-water reactor (LWR) demonstration plants.[9] The plant is owned and operated by a group of utilities. The construction contract for the plant was placed directly by the utility group, not by a government agency. Government funds for construction and operation of the plant are given to the utility group. Thus the supply contract is modeled after normal commercial contracts. According to common practice in the West German nuclear power plant business, the contract was to be of a turnkey type. The supplier company functions as main contractor and as architect-engineer for the whole plant, except for the civil engineering that is contracted to a consortium of constructors.[10]

The West German Science Ministry favored this organizational scheme in spite of a different suggestion from Karlsruhe. In 1965, the Karlsruhe laboratory had proposed forming a company made up jointly of government laboratories or agencies (Karlsruhe, the Belgian CEA, the Dutch TNO) and utilities from the three countries. This company would order the plant from producer industry, supervise construction, pursue licensing and regulation procedures, and hand over the plant after completion to the utilities for operation.[11] The Science Ministry held, however, that the role of the research laboratories should be reduced at the prototype stage. It preferred the scheme already proved in the three West German LWR demonstration plants, which gave much more responsibility to the utilities.[12]

The formation of the utility company to own and operate the SNR-300 was undertaken in two steps. A preliminary company Projektgesellschaft Schneller Brüter GbR (PSB) was founded on 21 November 1969 by Rheinisch-Westfälisches Elektrizitätswerk AG (RWE) of West Germany, Synatom S.A. of Belgium, and N.V. Samenwerkende Electriciteits-Productiebedrijven (SEP) of the Netherlands. The objective of this company was to arrange for the utilities all preconditions for construction and operation of the prototype plant in terms of finance, contracts, and licensing. In accordance with the preliminary nature of the company, a legal form was chosen which did not grant it a legal personality. The Schnell-Brüter-Kernkraftwerksgesellschaft (SBK), which was the legal personality to enter into the contracts as prepared by PSB and actually to construct and operate the prototype plant, was founded on 27 January 1972. Negotiations for construction of SNR-300 were not yet completed at this time. But the formation of this company was necessary in order to place the first part-orders in March 1972.[13]

The three utility partners are not of the same type. SEP is an association comprising all Dutch utilities. Synatom is a joint subsidiary of several Belgium utilities. RWE is the largest utility in West Germany, providing a third of West German electricity supply. After a change in its company objectives, in 1977 Synatom transferred its shares to Electronucléaire, which is a joint subsidiary of almost the same utilities.

For SBK, a total capital stock of DM120 million was envisaged, of which RWE was to contribute DM84 million, Synatom and SEP each DM18 million. In May 1973, after start of construction, the British Central Electricity Generating Board (CEGB) entered as a fourth partner with a share of DM2 million.

As contractor on the supplier side, the Konsortium SNR was reorganized in 1972 as Internationale Natrium-Brutreaktor-Bau-Gesellschaft mbH (INB). This company has a stock capital of DM0.1 million, of which Interatom holds 70 percent, Belgonucléaire and Neratoom each 15 percent.[14]

The whole organizational structure for the construction and operation of the SNR-300 plant is tied together by a number of contracts or legally binding statements, of which the more important are (see figure 6-1):

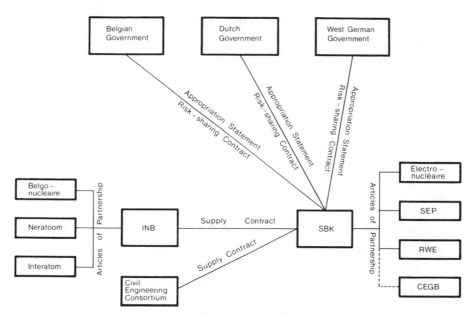

Figure 6-1. Contractual Framework for the SNR-300 Plant

The supply contract between SBK and INB

The supply contract between SBK and the civil engineering consortium

The fuel fabrication contract of INB with Alkem and Belgonucléaire

The appropriation statements by the three governments

The risk-sharing contracts between the three governments and SBK

The articles of partnership for SBK, which were to be agreed by RWE, Synatom (later Electronucléaire), and SEP

The articles of partnership for INB, which were to be agreed by Interatom, Belgonucléaire, and Neratoom

All these agreements are interrelated and had to be negotiated before final decisions could be taken. The governments played a decisive role in their negotiation, as they cover most of the expenses. Even the contracts that did not directly involve the governments needed their approval. Most of the negotiating was done by a group set up specially for this purpose in the West German Science Ministry at Bonn.

Like normal commercial contracts, the SNR-300 contracts include some guarantees associated with conventional penalties in case the plant should not come up to its specifications. Exempt from the usual guarantees are parts for which, because of their prototype nature or because of uncertainties in the outcome of the associated R&D program, the risks are beyond the usual level in power-plant construction. Nevertheless, the contracted guarantees include an obligation of the supplier to construct a fossil-fueled steam-supply system in case the nuclear steam-supply system should not reach a certain output within a specified time after startup. The utilities emphasized that all the obligations of INB must be guaranteed by the firm's parent companies and liability for Interatom's contractual obligations must be taken over by Kraftwerk Union.

While government funds for construction and operation of the SNR-300 plant are given to SBK, the funds for associated R&D performed by the supplier consortium and by the government laboratories are appropriated directly to the performing institution, each government paying for the activities performed by the organizations in its country.

In the course of the negotiations on the SNR-300, Luxemburg pulled out of the project. Belgium and the Netherlands increased their financial contribution from 12.5 to 15 percent, and their shares of supplies were redefined. The Netherlands was now to supply 5 percent of the plutonium and all sodium components. Belgium was to provide 40 percent of the remaining plutonium and a similar share of the fuel assemblies.[15]

Licensing Requirements

The organizational scheme for the SNR-300 implies that this plant is subject to the same licensing procedure as are commercial reactors. Thus the SNR-300 incorporates the safety features of possible future commercial breeders. In this regard it is different from similar prototypes in Britain and France which were built under special licensing procedures under the authority and responsibility of the state agency carrying out the R&D.

In July 1970, the Reactor Safety Commission put forward a general policy about the siting of prototype plants. This precluded construction of the SNR-300 on a previously planned site near Weisweiler, since this area was too densely populated according to the new criteria.[16] The necessity to look for a different site did not delay construction of the plant, as there were lags in the associated R&D program, and as the licensing authorities demanded drastic alterations in the plant design that required a good deal of time.[17]

The most severe alterations related to hypothetical accidents. The safety philosophy of the original design placed emphasis on measures to prevent a nuclear excursion. It was thought that this accident could be made sufficiently unlikely to avoid measures accommodating its consequences. The licensing authorities, however, not only required more stringent measures for prevention of this accident, but also stipulated that the impact of a nuclear excursion should be safely accommodated by the reactor vessel and the containment. They specified that the reactor vessel should be designed to accommodate a power burst up to 150 megawatt-seconds (MWsec).[18] The hypothetical case that the reactor vessel would not remain intact was to be provided for by designing the containment housing the primary coolant circuit to withstand a power burst up to 370 MWsec. In addition, the hypothetical case of a collapsed reactor core melting through the reactor vessel was to be provided for by installing a core catcher, a device which can safely receive molten fuel, prevent it from forming a critical configuration, and cool it sufficiently to avoid damage to the containment.

A redesign of the primary coolant circuit was required to guard against the hypothetical case of loss of coolant through breakage of one of the main coolant pipes. The components of the primary coolant circuit such as the reactor vessel, pumps, and intermediate heat exchangers had to be arranged in a way so that a minimum sodium level is maintained whatever breakages in the piping or the components occur. With this minimum sodium level, the decay heat from the core can be removed, either by a coolant loop, as long as one remains intact, or by an emergency cooling system. To eliminate possible sodium voids through gas bubbles entering the reactor core, a special device had to be designed in the coolant inlet of the reactor vessel which separates the gas from the sodium before the coolant passes through the core.

Other measures stipulated by the licensing authorities were not related to the specific safety features of the liquid-metal fast breeder reactor (LMFBR), but resulted from a general tightening of standards on the safety and the environmental impact of nuclear reactors. To decrease the thermal impact of the cooling water on the Rhine river, a cooling tower had to be built. The specifications for the reactor containment to withstand airplane crashes were increased. Whereas the initial design provided for a crash of a *Starfighter*-type plane, now the containment had to withstand the impact of a *Phantom*-type plane. Furthermore, the explosion of a gas cloud on the Rhine river was added to the list of external impacts the plant had to withstand. Substantial alterations were also required in the emergency cooling system, and additional safeguards were demanded for excluding access by saboteurs.

The demands of the licensing authorities substantially increased construction costs and delayed construction start by about one year. In addition, they necessitated a lot of additional R&D in industry and government laboratories. This added considerably to the total costs of the fast breeder program. For some items, the licensing requirements even went to the limits of technical feasibility, and imposed heavy tasks on scientists and engineers in industry and government laboratories.

Adapting Design to User Needs

Larger prototype plants are frequently called demonstration plants. However, it is often not clearly stated what is to be demonstrated. West German utilities distinguish between two types of demonstration: demonstration of commercially usable power generation, and demonstration of commercial power generation. As a utility must supply electricity at the very moment when it is demanded, it places utmost emphasis on reliability in power-generation plants. For the electricity produced by a plant to be commercially usable, the plant must fullful certain specifications for availability and length of operation cycles between shutdowns for inspection, maintenance, and refueling. Demonstration of commercial power generation in addition requires competitive costs of electricity.[19]

The SNR-300 was to demonstrate that this design of an LMFBR is capable of producing power in a manner so it is commercially usable. Although it did not have to show competitive power costs, it was to provide a perspective on the cost potential of larger plants of this type in the range of 1,000 to 2,000 MWe. A 300-MWe prototype allows better extrapolations, the more it uses a plant layout and components similar to those expected for large plants.

It was with a view to these two factors (production of commercially usable power and cost extrapolation to large fast breeder plants) that the utilities demanded substantial alterations in the SNR-300 design.[20] The original design evisaged a cylindrical containment, which the utilities wanted replaced by a cubic one. Such a design, which is possible in liquid-metal fast breeders because of low design pressure, provides much more space for handling fuel elements and for repair and exchange of components. Utilities emphasize easy handling of fuel elements as well as easy repair and exchange of components in order to keep outage times for refueling and repairs at a minimum. The original design was criticized particularly because the steam generators, which are the biggest single components posing specific technical problems, could not be exchanged at all, and their repair had to be done in difficult working conditions.

For reload of fuel elements, the original plans envisaged an interval of three to four months. As the plant has to be shut down about three weeks each time, the utilities considered this interval too short. They pressed for a half-year interval, so that one reload each year could be done in the summer, when demand for electricity is low in Germany.

According to the original plans, the spent fuel elements had to be stored in the reactor tank for some time until their decay heat and radio-activity had decreased to such an extent that they could be taken out. But the utilities deemed it necessary to be able to take fuel elements out without delay in case of failures or for inspection. As a consequence, a new fuel-shuffling device had to be developed that was capable of managing a higher decay heat. The utilities also demanded that the fuel elements supplied by the West German and the Belgian manufacturers have identical technical specifications, since small differences in the proposed specifications would have complicated operation.

Another alteration related to the steam generators. According to the initial design, they were to be a straight-tube type as developed by Ner-atoom. As the utilities had some reservations on this type because of spe-cific technical features, they demanded that a different type, the helical-tube steam generator, should also be tested in the SNR-300. Originally, the helical-tube steam generator had been developed in West Germany, but it had been dropped when it was agreed later that all sodium components were to be supplied by Dutch firms. To reduce the technical risks and to establish an element of competition, the utilities would have liked to have the two types of steam generators supplied by two different companies. In their negotiations with the consortium, the demand for a second supplier was not accepted, but a compromise was reached that the Dutch supplier develop and supply both types. In at least one of the three coolant circuits of the SNR-300 plant, steam generators of the helical-tube type will be installed.

Because large steam generators promised to reduce costs and because the SNR-300 plant should allow as much extrapolation to larger plants as possible, the utilities demanded larger steam generators than originally planned. However, they did not insist further on this, when the supplier pointed out that larger steam genserators would require additional R&D and thus delay construction of the plant. Now the plant is being constructed with three steam generators in each of the three coolant loops, as originally planned.

A drastic alteration in the core design was demanded by the utilities with a view to reducing operating costs of the SNR-300. Fuel fabricating costs as quoted by the suppliers were so high that for the first twelve operat-ing years, a total loss of about DM500 million was estimated. The utilities therefore picked up a suggestion by the German fuel-element manufacturer to increase the diameter of the fuel pin from 6.0 to 7.5 millimeters, resulting in a reduction in fuel fabrication costs of 20 percent. To accommodate the associated reduction in power density, the core was enlarged and the radial blanket accordingly reduced. As the transition to a different fuel-pin dia-meter required additional development work and irradiation tests, the results of which will not be available in time, the reactor is to be started with 6-mm pins. A core with 7.5-mm pins will be installed after two reloads, and a new licensing procedure will have to be opened for it. However, some savings are already achieved in the first core by reducing the radial blanket and enlarging the active core. Altogether these alterations reduced the expected loss for the first twelve years of operation by about DM200 million. This was at the expense of a reduction in the breeding ratio to a value around 1 or slightly less for the first core, and marginally above 1 for the subsequent cores. Therefore the SNR-300 will start to operate as a breeder reactor that does not breed.[21]

In the Karlsruhe laboratory where the plant had originally been set the objective to demonstrate breeding,[22] this decision was not received with sympathy. The scientific council of the laboratory passed a motion criticizing it. However, there is no technical or economic need to operate the SNR-300 plant at the highest possible breeding ratio. The potential of this design for breeding can be easily calculated within a small uncertainty band, once the actual breeding performance of an SNR-300 core is measured, even if the measurements are done in a core designed for a low breeding ratio. In view of the savings in operating costs on the order of a few hundred million deutschmarks, the small uncertainty in this calculation is acceptable. It seems, therefore, that the objective to demonstrate breeding in the SNR-300 plant as previously set by the Karlsruhe laboratory was guided by technical fascination and prestige thinking rather than by principles of a cost-effective development strategy. Of course, the reduction in the breeding ratio affected adversely the public relations of the SNR-300 project. The Karlsruhe scientists valued public relations more highly and were prepared to spend more money on it than the utilities.

The larger fuel pin will probably be chosen also for the future FBRs with outputs of 1300 MWe or more. Thus the transition to the larger pin has rendered some of the earlier work on 6-mm pins obsolete. This shows that a thorough and realistic economic analysis is important for directing technical development work. Interviewees from the Karlsruhe laboratory pointed out that in the 1960s there was general agreement, in all countries active in fast breeder development, that the economic optimum was at a 6-mm diameter.[23] It was natural that Karlsruhe's own calculations, which leaned on the international literature, found the same optimum.[24] Producer industry seems either to have followed the specifications given by Karlsruhe without further economic analysis, or to have been reluctant to advance a new assessment and the resulting technical alterations. Thus, it needed the utilities to bring into the project an independent economic judgment and to change directions in fuel-element development.

Costs

On 8 December 1969 the consortium communicated to the ministry an informal price estimate of DM550 million (not including the core and owner's cost).[25] As ministry officials and industry managers deemed this too large a departure from earlier estimates, the consortium checked the design for opportunities of saving. When submitting the designs on 31 December 1969, it gave a reduced price estimate of DM517 million. This was still an informal estimate and was not communicated as a binding tender, though industry was obliged by the appropriation statements of 1966 to submit a fixed-price tender. Ministry officials and the management of the consortium agreed to wait with a binding offer until the inflationary push in capital goods at that time, the first to hit West Germany after a long period of relative monetary stability, would be over and better prices could be negotiated with subsuppliers.

The problem of the price lost its significance, however, when substantial design changes were demanded by the licensing authorities and by the utilities. Serious negotiations about the price did not seem possible before the licensing authorities had accepted the basic specifications of the new safety features, before the alterations demanded by the utilities were finalized, and all the resulting changes in the design had been worked out in detail.

The West German parliament and the public were alerted by published cost estimates for the American 300-MWe fast breeder prototype plant, quoting a figure of $400 million to $500 million (DM1,400 million to DM1,800 million).[26] Even if $100 million were deducted for R&D, this was more than double the German estimate as of 31 December 1969.

As work on the revised design for the SNR-300 proceeded, cost estimates escalated rapidly (see table 6-1). In July 1970 an informal figure of DM617 million was given, reflecting some of the design changes required by the licensing authorities and utilities. The first estimate to be released to the public was that of February 1971 amounting to DM670 million.

The utilities did not pay much attention to the early cost estimates, as they deemed the plans not yet detailed enough to make a contract. They regard estimates made before designs are ready for a contract as empty talk. However, it was still something of a shock when they were given the first tender by industry in April 1971 quoting a still not binding estimate of DM723 million. As much lower figures had been announced for the French prototype Phénix before,[27] a thorough comparison of SNR-300 and Phénix was done. This comparison convinced them, however, that the figures given by the consortium were not out of proportion.

When reviewing the price estimate of April 1971, the utilities found that the designs had not yet progressed in sufficient detail for an assessment of the price. They viewed about 80 percent of the expenditure items (not including the civil engineering) not detailed enough to allow judgment on the price worthiness of the offer. Experience with the price estimate for the civil engineering part made the utilities look skeptically at the whole estimate. While the Konsortium SNR in April 1971 rated civil engineering at

Table 6–1
Cost Estimates for the SNR-300 Prototype Plant
(in million deutschmarks at constant July 1972 prices, current prices in brackets)

No.	Date of Estimate	Source of Estimate	Plant Excluding Owner's Costs[a] and First Core	First-Core Fabrication	Plant Including First-Core Fabrication	Owner's Costs[c]	Plant Incl. First-Core Fabrication and Owners Costs
1.	July 1965	Karlsruhe	359[b](260)[b]		415[b](300)		
2.	October 1965	Karlsruhe		69[b](50)	428[b](310)		
3.	December 1965	Ministry/Atom Commission			620 (450)		
4.	May 1967	Karlsruhe/Interatom	352 (260)				
5.	8 Dec. 1969	Konsortium SNR	652 (550)				
6.	31 Dec. 1969	Konsortium SNR	613 (517)				
7.	1 July 1970	Konsortium SNR	699 (617)				
8.	11 Feb. 1971	Konsortium SNR	707 (670)	63 (60)	770 (730)		
9.	15 Apr. 1971 (first tender)	Konsortium SNR[c]				95[c](90)	865 (820)
10.	1 Oct. 1971	Konsortium SNR	758 (723)	63 (60)	821 (783)		
11.	7 Feb. 1972	Konsortium SNR	974 (942)	54 (52)	1,028 (994)		
12.	10 Nov. 1972 (binding tender)	Konsortium SNR			1,189 (1,177)		
	Supply contract[d]	SBK/INB	940	60	1,000	103	
	Provisions for additional costs[d]		232				
	Total		1,172		1,232	103	1,335
13.	Oct. 1975	SBK/INB	1,440	60[e]	1,500	142	1,642
14.	Dec. 1977	SBK	1,855	80	1,935	179	2,114
15.	Dec. 1978	SBK	1,899	115	2,014	199	2,213
16.	June 1979	SBK	2,117	135	2,252	200	2,452

Sources:

1. Working Group, appendix 5, minutes of 49th session, 2 July 1965.

2. Kernforschungszentrum Karlsruhe, Projekt Schneller Brüter, "Ausführliche Erläuterungen zu den Anträgen zur Bereitstellung der Mittel für die Erstellung der Unterlagen zum Bau der Prototypen des Schnellen Brüters," October 1965, p. 65.

3. Ad hoc committee of the Working Group, minutes of 1st session, 15 December 1965.

4. K. Gast and E. Schlechtendahl, "Schneller Natriumgekühler Reaktor Na2, Report KFK-660, EUR-3706-d (Karlsruhe: Gesellschaft für Kernforschung, October 1967), pp. 8-1 to 8-2.

5–12. Interviews with participants. The figures for owner's costs in estimate no. 8 have been taken from the statement of a ministry official in a public discussion; see "Wo steht die deutsche Schnellbrüterentwicklung? Die öffentliche Diskussion in Karlsruhe und der Statusbericht 1971," *Atomwirtschaft* 16 (1971):195–208.

13. A. Brandstetter and A.W. Eitz, "The Tripartite Fast Breeder Programme: A Utility/Industry View," *Nuclear Engineering International* (July 1976):40–43.

14–16. Interviews with participants, October 1978, June 1979, March 1980.

Note: Cost estimates prior to November 1972 were inflated to July 1972 prices using the price index for industrial capital goods. Because of the introduction of a new value added tax in 1968, price indexes before and after this time are not directly comparable. Some adjustments had to be made; it was seen that these overestimate rather than underestimate inflation.

[a] Owner's costs include personnel, licensing, land, tax during construction, consultancy, electric power for construction, spare parts and materials, and roads.

[b] The Karlsruhe estimates of July and September 1965 do not specify whether they cover owner's costs and the costs of plutonium.

[c] The figures for owner's costs were announced by a ministry official and were not necessarily estimates of the Konsortium SNR.

[d] The price agreed in the supply contract of November 1972 does not include reserves for contingencies (DM158 million), and additional costs of foreseeable licensing requirements (DM74 million). These items are listed as provisions for additional costs.

[e] Extrapolated from the estimate of November 1972.

DM100 million, a binding tender invited from an international consortium of civil engineering companies quoted a price nearly twice as high.

By March 1972 the utilities considered the design still not detailed enough to sign a contract. First part-orders were given on 23 March 1972 to the Konsortium SNR (DM16 million) and the civil engineering consortium (DM8 million) in order to finance continuation of the planning activities to the point where the price could be assessed and construction could start. Together with the first part-orders, letters of intent were issued to the two consortia, that is, the utilities obliged themselves to cover the costs of the design activities of the suppliers in case of a cancellation of the project. In practice, the first part-orders together with the letters of intent meant for the Konsortium SNR that government funds for the design activities were now channeled via the utilities, and not directly as before.

A major obstacle for the negotiations resulted from the consortium's refusal to quote a fixed price for all its supplies. The consortium was prepared to commit itself to a fixed-price tender only for the conventional parts, such as turbogenerator and switch gear, but not for the major part of the nuclear steam-supply system. This threatened to bring negotiations to a halt, as it would have implied an open-ended commitment which the federal budget law did not permit. The ministry kept the negotiations going, however, by suggesting a price model that was basically a cost price but provided penalties to the supplier for costs transcending a previously agreed limit. A schedule was worked out that the consortium would finance 10 percent from the first DM41 million, and the penalty was to increase with further cost-overruns so as to reach 50 percent after DM166 million.

Agreement on the main contract conditions was reached in December 1971. On this basis, the consortium submitted in February 1972 the final tender, quoting a price of DM1,177 million (including fabrication of the first core). In subsequent negotiations this price was reduced to DM1,000 million. This reduction was largely cosmetic, however. Part of it resulted from cuts on supplies. Another reduction was achieved by decreasing penalties to the supplier for cost-overruns in the supplies at cost price. The government agreed to finance 100 percent of cost-overruns up to DM35 million, 95 percent of the next DM35 million, a share of 50 percent being reached after DM145 million.[28]

The DM1,000 million agreed on in November 1972 did not include owner's cost, costs of plutonium, and inflation. Furthermore, provisions had to be made for the governments' share in the cost-overruns and for additional costs of foreseeable licensing requirements. If these items are added, total costs as estimated in November 1972 amounted to DM1,588 million (see table 6–2).

The DM158 million provided for the governments' share in possible cost-overruns represent a ceiling. The whole contract will have to be rene-

Table 6–2
Cost Structure of the SNR-300 Plant, as of November 1972

	Million DM
First part-order to INB	16
Supplies at fixed price	415
Supplies at producer costs	290
Operation startup	22
Engineering	8
Allowance for lack of detailed design	15
Civil engineering	174
Fabrication of first core	60
Total orders (excluding plutonium)	1,000
Owners costs (licensing, land, personnel, tax)	103
Additional costs of foreseeable licensing requirements	74
Reserves for contingencies (governments' share of cost-overruns)	158
Total in 1972 prices (excluding plutonium)	1,335
Provisions for inflation	200
Total in current prices (excluding plutonium)	1,535
Plutonium	53
Total	1,588

Source: Interviews with participants, December 1974. The estimate for the costs of plutonium was derived from published sources, assuming a 57-percent West German contribution; see Federal Ministry for Research and Technology, *Fourth Nuclear Program 1973 to 1976 of the Federal Republic of Germany* (Bonn: Ministry of Research and Technology, 1974), p. 39.

gotiated, if this amount is reached. According to the practice adopted by the heavy-plant business in the early 1970s, inflation is provided for by escalation of the price according to certain agreed price indexes. Published figures on the costs of the SNR-300 plant included allowances for inflation which followed the official projections of the West German government. The official estimate of inflation in 1972 assumed 3 percent per annum for material and 7 percent for manpower. Given the initial construction schedule and the initial cost volume of the SNR-300 plant, this made a total of DM200 million.

The major part of the cost increase from 1970 to 1972 was real. It did not result from inflation or from the costs of additional safety features. In order to exclude the effects of inflation, the figures in table 6-1 use a common price basis as of July 1972. The effect of additional safety features was estimated by interviewees at DM60 million to DM80 million in 1972 prices. Thus there was a real increase of several hundred million deutschmarks that reflects the difference between an informal estimate and binding negotiations.

The considerable increases in the price of the SNR-300 in 1971–1972 from one estimate of the consortium to the other strained the ministry's relations with Interatom. Whereas ministry officials in 1970 were indulgent to the consortium not submitting a binding tender, the high increases in the price estimates in 1971–1972 seemed incomprehensible to them. Questions were raised as to whether the supplier company was trying to make a hidden profit on the SNR-300 contract. A review of the tenders by the ministry's accountants, however, reassured the ministry that the tenders were calculated in an orderly manner. Finally it had to accept the fact that the cost increases were real.

After construction start of the SNR-300, costs continued to escalate. As of June 1979, the utility estimates a total of DM2,452 million in 1972 money (including first-core fabrication and owner's cost). A large share of the DM1,117 million difference to the 1972 estimate is due to the requirements of the licensing authorities; the DM74 million provided initially for this was exhausted in 1974. Part of the cost increases stem from the delay of construction, which caused some underutilization of the supplier's personnel. Another part arises, in the utility's view, from real cost-overruns in terms of the contract, and the ceiling for the governments' share is nearly reached. More important, it has increasingly become impossible to determine what share of the cost-overruns results from licensing requirements and what is real, as the licensing requirements changed the specifications of most components. This determination will finally have to be made not by accounting but by negotiating. In other words, the cost structure worked out for the SNR-300 plant has been made inoperational by the licensing requirements.

Even ignoring the cost increases during construction, the cost information in the negotiations of 1970 to 1972 showed previous estimates to have been unrealistic (table 6-1). The first estimate by the Karlsruhe project management in July 1965 gave a figure of DM300 million (including first-core fabrication). In October 1965 the project management came up with a more cautious estimate, as the amount of DM310 million (with first-core fabrication) was not meant to include all the plant components; some of the sodium components from the test rigs were assumed to be installed in the plant. Ministry officials were skeptical of the Karlsruhe estimates. Their estimate of DM450 million in December 1965 was adopted by an ad hoc committee of the Working Group.

The Karlsruhe laboratory rejected this skepticism, repeating in 1967 its former estimate (DM260 million without core) in a study done in close collaboration with Interatom. This time the use of components of the test rigs in the plant was not assumed, and all the plant components were included. The study expressed pronounced confidence in the estimate, stating that it was based on conservative assumptions with regard to safety-related design features and that prices were estimated on a first-of-its-kind basis:

On the whole, the capital costs as estimated at about 260,000,000 DM can be regarded as a reliable figure. In a comparison of investment costs with corresponding figures for thermal nuclear power plants of the same rating, it has to be remembered that this study deals with a prototype plant, and components of the reactor and of cooling circuits with very few exceptions are fabricated for the first time and as single pieces. Furthermore the conservative assumptions with regard to reactor safety as made in this study have the effect of cost increases. Further investigations will show which savings will be possible in future.[29]

As of 1972 costs were about three times higher, at constant prices. By 1979, they had reached a level six times higher, at constant prices.

Risk-Sharing Contract

A critical focus of the negotiations was the contribution of the utilities to investment costs and their share in possible future operating losses.[30] Both points are directly interrelated. If a prototype power plant did produce electricity at competitive cost, it could be financed by a utility, without government aid, in the usual way: 30 to 40 percent would be self-financed, the rest would be covered by bank loans. The returns from the sale of electricity would be sufficient to pay for interest and repayment of the loans as well as for depreciation and the usual profit on the own capital. The nearer a prototype plant comes to competitive costs, the larger is the commercial part of the construction costs, that is, the proportion that can be recovered from the sale of electricity and thus can be financed in the usual commercial way, and the smaller is the uncommercial part to be covered through subsidies, either from the utility itself or from the government.

Because many new technical features are tried for the first time in a prototype plant, there is a greater risk than usual that the plant will not fulfill its design specifications, or will suffer from breakdown. This risk can be underwritten by the state partly or totally through risk-sharing agreements.

The net revenues from operation of a prototype plant are a function of the price of the electricity sold, and of the costs of operation including fuel-cycle costs. The pricing of the electricity produced in a prototype plant is not a straightforward matter, as it has to take into account that supplies from a prototype plant cannot be relied upon because of the greater risk of forced outages. Unreliable electricity is of little commercial value, as standby capacity is necessary which can take over whenever a forced outage occurs. Because of the uncertainties involved, the pricing of electricity from prototype plants leaves some scope for negotiation between the utilities taking the electricity from the plant and the government subsidizing the plant. In the case of the SNR-300, this problem was settled by charging, up to an annual load factor of 72.5 percent, a price equivalent to the costs of electricity in commercial LWRs, and beyond this load factor a price equivalent to the variable fuel-cycle costs. The utilities emphasize that, in view of the small commercial value of the plant's electricity, the supply of which is unreliable and not fitted to electricity demand, this price arrangement implies a genuine financial contribution on their part.

Originally the utilities offered a financial contribution of DM100 million to the investment costs of the SNR-300. This was at that time about the amount which a utility had to invest from its own funds (excluding bank loans) in 300-MWe generating capacity of an LWR plant. The associated scheme of risk sharing provided that, in scheduled operation of the plant, the utilities could depreciate their capital of DM100 million within fifteen years and were credited a modest interest on this capital, but would not make any profit in addition to depreciation and interest on this capital. In the case of losses, the capital would be depreciated as scheduled. Losses would be covered by reducing interest. Further losses would be paid by the three governments up to a ceiling of DM100 million.

It came as a distressing surprise to all the organizations involved in the SNR-300 project, when estimates of future revenues from operation of the SNR-300 plant, calculated on the basis of this finance scheme and tenders for fuel fabrication submitted in November 1971, showed a total loss of about DM500 million within the first twelve operating years. The ceiling of DM100 million for government guarantees would have been reached within three years. It was evident that the SNR-300 project could not be implemented on this basis. If no means to reduce the expected loss could be found, the project was seriously threatened. When issuing the letter of intent in March 1972, the utilities made therefore the final order contingent upon a satisfactory solution of the fuel-cost problem.

In subsequent negotiations, a package of measures was worked out, through which the losses were brought down to an acceptable level. The most effective measure, saving about DM200 million, was found in technical alterations in the reactor core. For a further reduction in fuel-fabrication costs, the fuel manufacturers were protected by government guarantees against certain technical risks, so a new fuel-supply contract with lower prices could be worked out. Furthermore, the governments took over some insurance risks, so savings were made on insurance premiums. Finally, the utilities increased their capital from DM100 million to DM120 million, and agreed that DM20 million would be depreciated only when there were no losses to be financed by the government. Under the new scheme, the governments would have to bear in total no losses, assuming normal operation of the SNR-300. Their guarantee would be called on only temporarily up to a cumulative sum of DM42 million. The utilities would get in twenty-five years of operation a cumulative interest on DM100 million of their capital which, averaged over this period, would amount to less than 6 percent per year. In the final negotiations on the risk-sharing contract, the ceiling for government guarantees was raised to DM186 million. The plant is to be shut down when 80 percent of this sum is exhausted, so the remaining funds are available to cover the cost of decommissioning.

At the time when the contracts were finalized, the DM120 million capital stock of SBK corresponded to about 8 percent of total estimated construction costs (without plutonium). When construction costs grew in later years, the shareholders committed themselves to increase capital stocks of SBK proportionately.

Economic Evaluation

Advisory Committees

The ministry initially planned the evaluation of the SNR-300 project in two steps. First, the Fast Breeder Project Committee was to give a judgment on the technical maturity of the SNR design. Second, the Working Group of the German Atom Commission was to give a recommendation on whether government support should be given. Officials presented this proposal to the Working Group on 6 March 1970. At this session, however, the Working Group considered a positive statement on the technical maturity of the SNR-300 project by the project committee to be a precondition for an evaluation of the justification for government support and, therefore, deferred the establishment of a special subcommittee.[31]

In January 1971, though the project committee's consultations on the technical maturity were still underway, the ministry proposed to the Working Group that it undertake the economic evaluation in parallel. In particular, the ministry desired a recommendation on the following items:

1. Potential alternatives to FBRs for economic electricity generation in the long term.
2. The time scale for the deployment of economic fast breeders.
3. The optimal strategy in terms of time and cost for future development of FBRs up to commercial introduction.
4. The expedience of a construction start for the SNR-300 as of spring 1972.

A subcommittee was established by the Working Group to review these items. It comprised four members who were chosen from institutions not directly involved in the SNR-300 project. Three were from the nuclear sector, one from a university institute for energy economics.[32] After six meetings, the subcommittee agreed in September 1971 on a final report, which had been drafted by two ministry officials.[33] As alternatives to fast breeders, the subcommittee discussed fossil fuel, nuclear fusion, solar, wind, hydroelectric, and geothermal energy. Fossil fuels (including lignite, coal, oil, and gas) were assessed as having a resource base sufficient to cover eight to eighty times estimated cumulative world demand up to the year 2000, depending on whether only assured reserves or also additional probable resources were considered. However, problems were seen in the risk of long-term depletion and in possible cost increases due to environmental requirements. Fusion was considered a genuine long-term alternative to fast breeders, but too far off to be taken into account in present decisions in energy policy. Other sources such as solar, wind, hydroelectric, and geothermal were assessed as having a potential only of local significance.

As to the timing of commercial introduction of fast breeders, the committee agreed that it was reasonable to expect that low-cost world uranium resources were sufficient to cater for an expanding world electricity economy based on LWRs until the year 2000. Thus it was considered sufficient that economic fast breeders be available in the 1990s, as then there would be a relatively large risk of a price increase for natural uranium. The subcommittee stated that electricity costs competitive to LWRs could not be expected for fast breeders before that time, because of the high investment costs of fast breeders. With regard to the optimal development strategy the subcommittee recommended construction of two additional demonstration plants to follow the SNR-300. For a reduction of the technical and financial risks, overlaps in the construction schedules of these plants should be avoided.

The subcommittee discussed in detail to what extent costs could be reduced through international collaboration. It did not go so far as to consider the possibility of abandoning domestic manufacture and of importing fast breeder plants, if and when they were needed. Instead it explicitly stated that its reasonings were based on the presupposition that manufacture of FBRs by domestic firms in the West German–Benelux countries was desirable. It did not give a judgment as to whether this presupposition was economically justified.

An extreme option was to resign totally from a domestic development effort and to take out foreign licenses when an economic use of fast breeders appeared possible. The subcommittee recommended that the government drop this option, if it saw a great risk that foreign licenses then would not be available, and if it attributed to fast breeder development a positive signal effect for exports of other power plants. It refrained, however, from any judgment as to whether there was an actual risk that licenses would not be available, or whether fast breeder development did have a positive signal effect for export.

A less extreme option was seen in dropping the SNR-300 plant and seeking instead foreign collaboration for a 1,000-MWe plant to be constructed either in a foreign country or in West Germany. This option was regarded as unfavorable, as it would give domestic industry little influence on the establishment of future collaboration. If the 1,000-MWe plant was constructed in West Germany and ordered by West German utilities, the utilities in West Germany and the Benelux countries would have some influence. However, according to information submitted by the West German supplier company, France and Great Britain were not prepared to collaborate in such an arrangement with West German–Benelux firms on acceptable terms, that is, on an equal basis, without prior construction of the SNR-300 plant. Nevertheless, the subcommittee stated that such an arrangement would provide financial advantages for the government and that the companies would accept unfavorable terms for collaboration if construction of the SNR-300 were precluded through prior policy decisions.

Thus the option was left open for construction of the SNR-300. The subcommittee recommended establishing an exchange of know-how with foreign partners during construction of the 300-MWe prototype and seeking their collaboration for development and construction of the subsequent larger plant. As the possible partners for collaboration, Britain and France, were planning to start construction of fast breeder plants with ratings around 1,000 MWe in a few years, a quick start of the construction of the SNR-300 was deemed desirable in order to be able to negotiate collaboration on equal terms for the later phase.

The subcommittee's justification was far from enthusiastic. The economic need for the fast breeder hinged on two points: the environmental acceptability of fossil power, and the risk of a price increase for natural uranium toward the end of the century. The fast breeder was seen as an insurance against the risk of rising uranium prices. A need for this insurance existed only if one did not want to fall back on fossil energy. With regard to international collaboration, the subcommittee addressed extremely well all important questions. But at critical points it evaded an answer by basing its conclusions on explicit presuppositions without assessing them or by referring matters to policy choices to be taken by the government.

The Working Group discussed the subcommittee's report on 28 October 1971.[34] This was the last session of the committee, as the ministry had decided to replace the apparatus of the Atom Commission by an Expert Committee for Nuclear Research and Technology and a number of ad hoc advisory bodies. The Atom Commission had already been dissolved at its final session on 19 October, so the Working Group was formally out of existence when it gave its recommendation on the SNR-300 plant.[35]

The subcommittee's report had been circulated before the session, and the members of the Working Group had been requested to write a comment on it. Three days before the meeting, the project committee had given a positive judgment about the technical feasibility of the SNR-300 plant. When introducing the subcommittee's report to the Working Group, a speaker of the subcommittee pointed out that the report was almost totally based on evidence submitted by the supplier side. A review of the evidence and of the report was necessary, which the subcommittee was not able to do completely. The speaker, who was a utility representative, suggested, therefore, that such a review be done by the utilities which were to own and operate the plant, with particular reference to a cost assessment of larger plants by extrapolation from the prototype plant.

After further discussions, the Working Group approved the subcommittee's report and recommended to the ministry construction of the SNR-300 as early as possible. At the time this recommendation was made, only preliminary information on the cost of the SNR-300 was available. For construction costs, there were only nonbinding estimates, since the final tender was not submitted before February 1972. Also information on fuel fabrication was incomplete, as the tender was submitted in November 1971. Nevertheless, with this recommendation of the Working Group the evaluation of the SNR-300 project was completed. The decision proposal for government support of the project went the formal route from the reactor section of the ministry up the hierarchy to the minister, through the cabinet and the joint commission of the three governments into the federal budget, without further review.[36]

The Bundestag

The Bundestag Committee for Science and Education discussed the fast breeder program on several occasions. A public hearing was held on 17 December 1970 where nine individuals—representatives of three reactor manufacturing companies, three government laboratories, and two utilities—gave evidence. The agenda focused on the question of whether the Bundestag should appropriate funds for fast breeders as well as high-temperature gas-cooled reactors (HTGCRs) and fusion.[37] A senior civil servant in the Land Government of North-Rhine Westphalia criticized the fast breeder on safety grounds, and, by implication, made a case for the HTGCR under development at Jülich. The representative of the Jülich research laboratory proposed reduction of the domestic effort on fast breeders through closer international collaboration, in case funds should not be available for financing both fusion and fast breeders. However, the level of questions and answers was very general. This was probably the reason why the chairman of the Bundestag committee stated at the close of the hearing that the experts left the members of parliament "relatively, though not absolutely" alone with their problems.

For detailed consideration of the SNR-300 project, the Bundestag committee in April 1970 set up a Fast Breeder Subcommittee, which up to the end of the legislative period in summer 1972 met six times. The sessions served mainly as briefings by ministry officials on the state of the contract negotiations. In the aftermath of the 1969 decision to terminate industrial R&D on the steam-cooled fast breeder, a good deal of the subcommittee's discussions was devoted to the status of the different coolants. There was an implicit understanding that the decision against the steam-cooled fast breeder meant a decision for going into the phase of prototype construction for the sodium-cooled fast breeder. This was expressed in the chairman's résumé of the subcommittee's session on 12 March 1971. He stated that, according to the subcommittee's opinion, the Rubicon has been crossed in fast breeder development. This meant that the sodium-cooled fast breeder had a clear priority, and that the economics of this type should be demonstrated through construction and operation of the prototype.[38] These statements, however, show that the objective of the SNR-300 plant was not clear to the subcommittee. The prototype was to prove only technical feasibility and reliability, not commercial competitiveness or viability.

In April 1971, the subcommittee, accompanied by ministry officials, made a trip to the United States in order to inform itself on the status and prospects of the fast breeder. It visited several companies and government institutions involved in fast breeder development. The statements received on fast breeder development in general and on the West German program in particular reassured the subcommittee in its positive attitude toward the SNR-300 project.[39]

Apart from this study tour, the subcommittee drew exclusively on information provided by ministry officials. At its final session on 9 June 1972, the Scientific Service of the Bundestag proposed ordering an expertise from an independent engineering consultant. Ministry officials, however, pointed to their disappointing experience with engineering consultants in other reactor projects. In their view no information could be expected from such an expertise which was not already contained in the final report of the subcommittee of the Working Group of the German Atom Commission. Although several members of the subcommittee were in favor of an expertise by an engineering consultant, no formal decision was taken. Thus the proposal did not lead to any action.[40]

The Official View

An official justification of construction of the SNR-300 was given in a paper by U. Däunert and W.J. Schmidt-Küster, both civil servants in the energy division of the Federal Ministry for Research and Technology.[41] Däunert and Schmidt-Küster referred to projections of future electricity consumption in West Germany which assumed a doubling of electricity demand roughly every ten years. Given that the present cost advantages of LWRs over conventional fossil-fueled plants would remain, it could be assumed that by the year 2000 the share of nuclear plant on total electricity-generating capacity would increase to about 75 percent. As known world reserves of uranium would cover projected world demand only up to 1990, there was a risk that the price of uranium would rise toward the end of this century, even if further reserves were discovered and better methods for prospecting and processing allowed mining of lower-grade ores without increasing costs. Fossil fuel, though it could cover energy demand until 2000 and long beyond, would imply the risks of long-term scarcity, increased costs due to environmental regulations, and political problems. As nuclear fusion had not yet proved technically feasible and as the HTGCR, though having better performance in the conservation of uranium reserves than the LWR, faced prospects similar to the LWR with regard to long-term uranium supply, the situation could be improved substantially only by developing fast breeders. FBRs could remove the risk of rising uranium prices by the 1990s, as by then they could be expected to have the same electricity costs as LWRs.

Construction of the SNR-300 was justified by Däunert and Schmidt-Küster specifically on the grounds that this plant would prove the advanced technical capability of the power-plant industry in West Germany and the associated countries: "The export of power plants plays an important role in world trade, and German industry with its Benelux partners must prove that it can master the advanced technology of the fast breeder (this proof being given in Britain and France through construction of the prototype-plants PFR and Phénix)."[42] This statement is rather ambiguous with regard to the export of FBRs. It attributes to this technology a signal effect for the high standard of domestic industry rather than a clear export potential.

Another argument related to future collaboration with Britain, which was eagerly sought by West Germany and its partner countries at this time. This collaboration seemed impossible without the three countries bringing into it the experience gained in construction of the SNR-300. The authors emphasized that the SNR-300, being a loop-type fast breeder in contrast to the PFR and Phénix being pool-type breeders, broadened the base of European experience in fast breeders. Finally, they considered licensing of a big fast breeder plant almost impossible without first constructing a smaller plant in the importing country.

It may be noted that this justification is a little more pessimistic on uranium availability than that by the ministry's advisory committee. More important, the justifications for the development of the fast breeder in West Germany changed totally over a period of two years. Until 1969, the main justification had been derived from the breeder's supposed short-term advantage in electricity costs over the LWR. As a result of the experience with the costs of the SNR-300, this assumption had become untenable. Even under optimistic assumptions, this cost experience, preliminary as it was, ruled out any possibility that FBR plants attain lower electricity costs than LWRs, unless uranium prices increased. Therefore, the 1990s became the new target date for the commercial introduction of fast breeder plants. It was assumed that by then the depletion of low-cost uranium reserves would have set in, and the breeder would be helped against the LWR by rising uranium prices.

Timing and Development Strategy

The timing of a prototype plant has to take account of two factors: the state of the development activities, on the results of which the plant draws; and the time scale of market needs. If a market is seen only in the long run, there is more time for reducing technical risk through tests and experiments before a larger prototype is constructed. If, however, competitors are likely to come on the market very quickly, it may be advisable to take higher technical risks in order to come on the market at the same time as the competitor, or not much later. As far as the SNR-300 plant is concerned, market prospects did not call for a quick start of construction. The price estimates made in December 1969, and even more those revealed in the subsequent negotiations, made it clear that the fast breeder would be commercially competitive only in the long run.

The SNR-300 substantiates the view that shortening the development time and taking high risks will result in higher development costs. Proponents of the fast breeder hoped that this technology would achieve its first commerical use by the 1990s, so a number of uncommerical plants would have to be built if industry was to be kept alive over this period. There was a case for delaying the SNR-300 plant as long as possible in order to reduce the costs of bridging this uneconomic period. A later start of construction would have allowed the reactor to start with the larger fuel pins. This would have reduced operating costs. A delay would also have allowed some savings in plant-investment costs, since the plant layout could have been optimized with regard to the alterations demanded by the licensing authorities. Furthermore, it would have allowed development of larger steam generators. This would have increased the amount of information useful for extrapolation to larger plants.

With regard to component testing, a delay of the SNR-300 plant would have allowed a sounder technological basis and would have reduced technical risks. The major test facilities for sodium components started operation in 1971.[43] The initial schedule for a construction start in April 1971 implied that there would be practically no scope for major design alterations between testing and fabrication. No time would have been available to incorporate major improvements into the components, and any severe failure of the components in the tests would have necessitated a construction delay.

There was very limited experience in fuel-element testing. By 1971, irradiation of forty-two oxide-fuel pins in the DFR was finished, irradiation of nine more in DFR and seventy-three in Rhapsodie was running.[44] This compares to thousands of fuel pins the British and French had irradiated in their small reactors, DFR and Rapsodie, before the fuel for the larger prototype reactors, PFR and Phénix, was fabricated.

Deferment of the SNR-300 for a number of years would have permitted use of the KNK reactor in order to broaden experience in fuel irradiation and sodium technology. By the time the thermal version of the KNK plant had started full operation (February 1974), the major components for the SNR-300 such as steam generators, intermediate heat exchangers, reactor vessel, and pumps had already been ordered.[45] This meant that the KNK plant could yield experience useful for SNR-300 only insofar as failures were detected during the operation startup and their causes identified. At that date, the KNK plant could not yet provide positive proof that the components would fulfill their specifications in the longer run. By the time the fast irradiations in the KNK plant yielded significant experience, the specifications for the SNR-300 fuel pins were finalized and fabrication was under way.

As far as the fast breeder fuel cycle was concerned, the status of reprocessing and refabrication left scope for great technical and economic uncertainties. Reprocessing of spent fuel for LWRs was (and at the time of writing still is) not yet done in a truly commercial type of operation. In Belgium and Germany fabrication of plutonium fuels was limited to small pilot facilities. Larger plants with a capacity of about 5 tonnes fast reactor fuel per year went into operation in 1973, after the decision to construct the SNR-300.[46]

These observations clearly establish the existence of a positive time-cost trade-off for the SNR-300 plant, though it is difficult to give exact figures for potential savings on cost. How was this trade-off perceived by the various institutions? What was their attitude toward the time schedule of the SNR-300 and the risks it implied?

Karlsruhe

The Karlsruhe project management approached the technical and economic risks of the SNR-300 project with an optimism that was tantamount to negligence. This can be grasped from a "possible schedule" of future fast

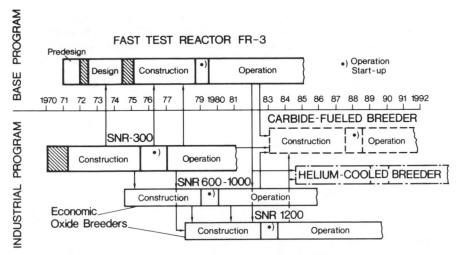

Source: Adapted from P. Engelmann, "Zukünftige Entwicklungsaufgaben des Projekts Schneller Brüter," *Atomwirtschaft* 15 (1970):334–336. Reprinted with permission.

Figure 6–2. A Possible Program Schedule as Proposed by Karlsruhe in 1970

breeder development outlined in spring 1970 (see figure 6–2). This schedule assumed a construction start of the SNR-300 in spring 1971, of a follow-up plant in the 600–1,000 MWe range in 1974, of a large fast test reactor FR3 in 1975, of a 1,200-MWe fast breeder power plant in 1978, of a carbide-fuel liquid-metal fast breeder in 1983, and of a gas-cooled fast breeder in 1984. The degree of optimism is illustrated by the project management's expectation that the 600–1,000-MWe plant would be economic or would have to a very large extent economic characteristics. The whole program plan and in particular the proposal for the 600–1,000-MWe plant must be seen as a last attempt by the project management to keep to the earlier formulated goal of the program: to have the first commercial fast breeder in operation by 1980.

The soaring costs of the SNR-300 plant soon proved this gigantic program to be unrealistic. At the status report of February 1971, the 600–1,000-MWe plant was already forgotten, and the FR3 received only passing mention.[47] The project manager's presentation focused on the need to proceed quickly with the SNR-300, because of the threat of international competition. Although the Americans had fallen behind the British and French, they were still regarded as the main competitors:

> We have seen and we still see, above all, American competition, the American challenge, to be decisive. . . . Like in the years 1965/66, the American competition is today still the determining factor. To the best of our present knowledge, in the United States construction will start in 1972 of at least one, probably of several liquid-metal prototype reactors. To the best of our present knowledge, this date is valid also for the SNR plant. We

are neck and neck with the partners which we regard as the main competitors, quite precisely neck and neck, although in 1960, i.e. ten years ago,
we were depressingly lagging behind. But it is to be remembered that the
American program is much broader (including the construction of the
FFTF). Further it is to be remembered that the British and French competition has a lead and is technically remarkable. It follows from this that the
decision to construct the SNR-300 must be taken by the end of this year.[48]

In retrospect, this impression of American competition has proved to be
unjustified. American plans for liquid-metal fast breeder prototype plants
have been reduced from three to one, and this one was delayed considerably
even before the U.S. government changed in 1977 its policy toward fast
breeder commercialization.

Supplier Industry

Speakers of supplier industry exercised constraint in public statements,
though having a vital interest in a timely start of construction. About half
of Interatom's activities was devoted to the SNR-300 design, and the design
contract ended in December 1969. Even if construction had started at the
earliest possible date, the company would have had to seek additional contracts from the ministry in order to bridge the time until construction start.
In addition to the design contracts of 1966, the company received other
contracts worth DM18 million in 1969 for work on integrated steam generators, on carbon-dioxide secondary cooling circuits, and on engineered safeguards. At the end of 1969, the earliest date of construction start appeared
to be April 1971, and the design contract was extended by an interim
program for this period worth DM28 million. The company was helped
over the further delay of two years by the ministry's allocation of funds for
the R&D program related to construction of the SNR-300 as early as 1971,
that is, a year before the final decision to build was taken and two years
before construction actually started. As from March 1972 continuing work
on the design was financed through the first-part order placed by SBK.[49]

The Ministry

Ministry officials felt that a delay of the SNR-300 plant would not decrease
the overall costs of the project, as continuing government support to Interatom was required in any case until construction start. Also budgetary considerations called for a timely construction start. Ministry officials anticipated that the ministry's budget would grow less in future years. While at
prevailing inflation rates every year of delay added to the price of the

SNR-300 plant, it would increasingly become difficult to accommodate the required funds within the ministry's budget.

A much stronger motivation for a quick construction start was their fear that a delay could have disadvantageous effects on the manufacturers. Ministry officials saw a general tendency for manufacturing firms dependent on state support to adopt more and more the behavior of a government research laboratory. This tendency would increase if construction of hardware was delayed and the teams were occupied by paperwork. Furthermore, the manufacturer indicated to the ministry that the most able scientists and engineers would rather leave the team than occupy themselves with paperwork. These undesirable effects could be avoided only by construction activities, in a contractual framework and this as near as possible to a commercial contract. In the absence of other major construction projects in the fast breeder program, ministry officials therefore thought that construction of the SNR-300 must start within a short time if the industrial fast breeder teams were to be kept together.

Although these considerations are a good illustration of organizational momentum in large-scale government-financed R&D, some perspective must be added as to their actual weight in policymaking. In 1969, the ministry admitted that Siemens practically dissolved its fast breeder team, when the company acquired a majority holding in Interatom and transferred its fast breeder activities to this firm. Siemens gave the engineers of its fast breeder team the option of joining Interatom at Bensberg or accepting work in other departments of the company at Erlangen. About a third of the team opted for Interatom.[50]

The ministry was less aware of cost uncertainties and of technical risks than the utilities. At a public panel discussion in February 1971 a ministry official declared:

> I think we should endeavor to make clear that the German fast breeder project today has reached a degree of detail that it is possible to state that we will finish this project for DM1,000 million at 1970 price level; and if we do not do that, then the firm Interatom, given for instance an increase to DM2,000 million, would have to pay up to DM500 million from its own funds. But the conviction of the experts sitting here—and this is the important point—must be such that they say we have a thorough grasp of the costs.

With regard to the much higher cost estimates for the American 300-MWe prototype plant, he added: "if the Americans build it at twice these costs, this is simply a fault of the Americans, but not an unhealthy excrescence of this project into the future.[51] A few months later the utilities stated that the price of 80 percent of expenditure items could not be assessed because designs were not detailed enough.

In the advisory bodies of the ministry, that is, the Fast Breeder Project Committee and the Working Group of the Atom Commission, discussions on the construction schedule centered mainly on the state of R&D activities and on questions of future collaboration. As mentioned earlier, prospects of future international collaboration were drawn upon to make a case for a quick construction start. R&D activities, it seems, were discussed in a tactical rather than strategic perspective. The consultations were preoccupied with the earliest possible construction start permitted by the state of R&D activities. In the design contract of 1966, the ministry had decreed that government money for construction of the SNR-300 plant would be allocated only after certain technical objectives were achieved in prior R&D. These objectives were formally framed as "appropriation conditions." It was on these appropriation conditions that advisory bodies' consultations were focused. Although the schedule for commercial introduction had been delayed by about two decades against expectations at the time when the appropriation conditions were formulated, they were accepted as terms of reference for the SNR-300 decision.

The more important of these appropriation conditions concerned test irradiations of fuel pins up to a certain burnup, operation startup of the KNK plant, and long-term tests of pumps, steam generators, and intermediate heat exchangers. The requirement for fuel irradiations was fulfilled with completion of two experiments in the DFR reactor in 1970. Other appropriation conditions needed a benign interpretation in order to allow a timely start of construction for the SNR-300. The ministry was prepared to admit this benign interpretation, though the appropriation conditions had been intended in 1966 to forestall pressures from the Karlsruhe project management for a premature start of construction, and had served this function well up to then. For example, in the consultations of the Working Group, the question was raised as to how the successful startup of KNK stipulated in the appropriation conditions should be defined. The ministry conceded that 40 percent of nominal output could be accepted as the criterion.[52]

As for other appropriation conditions, the decision to construct the SNR-300 plant was taken not upon their fulfillment, but on the assumption that these conditions would be fulfilled in the year in which construction would actually start. For instance, the long-term tests of the sodium components started in 1971 and then could be expected to be finished within about a year. Long delays occurred, however, in the steam-generator tests, which were completed in summer 1974 for the straight-tube and in the autumn 1975 for the helical-tube type.[53]

The ministry would have been in the position to delay the construction start of the SNR-300 plant by one or two more years simply by insisting on a rigorous interpretation of some of these conditions, especially those relating to the KNK plant or the testing of steam generators. However, it preferred to allow a flexible interpretation, as in its view some construction activities had to start within the next few years if the industrial teams were to be kept together.

The Utilities

The licensing authorities and the utilities were the only actors to urge some delay of the plant. The delay in the construction start of SNR-300 from April 1971 to April 1972 was due to licensing requirements. The utilities did not regret this delay, as it provided time to implement the design changes which they deemed necessary. It relieved them of the need to make up their mind whether or not to press hard for the design changes they wanted.

By contrast, the delay from April 1972 to April 1973 was forced by the utilities, because they were unwilling to take technical and commercial risks which they deemed unacceptable. They saw that important results of the R&D program would be produced so late that no leeway would be left for component fabrication. This meant that, in the case of unexpected results, specifications of ordered components would have to be altered. Specifically, this related to the startup of the KNK plant, which was to provide experience in sodium technology. As mentioned earlier, the utilities found that the designs were not yet detailed enough, so a price assessment was not possible for a good number of expenditure items. They concluded that these bottlenecks in the schedule unnecessarily increased the risks. A delay of about a year was therefore deemed imperative in their own interest as well as in the interest of the extraordinarily large public funds. Internally they considered even a more serious delay. They discussed a proposal to defer the start of construction for a couple of years and build instead a 600-MWc plant in collaboration with the British. No suggestion to this effect was, however, made to the governments or the advisory bodies, after the governments had agreed to reduce operating costs of the SNR-300 through further subsidies.[54]

International Collaboration

The chronology of events in the establishment of West German–Belgian–Dutch collaboration shows clearly that the initiative was taken by the research laboratories, and that in research laboratories and government agencies there was greater interest in this collaboration than in industry and utilities. Between the research laboratories the collaboration was formally established in 1965, while the governments followed in 1967, industry in 1968, and the utilities in 1969.

The motivations for this collaboration on the part of the government laboratories and of the West German Science Ministry were discussed in chapters 3 and 4. As for industry, West German firms did not like very much the idea of sharing the work with Belgian and Dutch companies, but once Karlsruhe and the ministry had taken the lead, they followed suit without much resistance.[55] West German know-how on sodium technology was ahead of that in the Netherlands, and hence this aspect of collaboration was not considered very attractive. Belgonucléaire, however, because of its contributions to the Enrico Fermi reactor in the United States, had know-how in fuel-element technology from which the Germans could profit.

The firms' reluctance to collaborate did not result from a high esteem for possible economic returns from the FBR but from their sensitivity of the frictions involved. In this regard, it does not matter whether collaboration is with foreign or domestic partners. Firms in general see consortia or joint ventures as appropriate only for clearly defined tasks. By definition, in a development project tasks cannot be sharply separated, as specifications for one component are interrelated with those for many other components.

Experience in the fast breeder program has confirmed such apprehensions. Friction with Interatom was one of the motivations for Siemens to take over this company. Also the later collaboration of West German firms with Belgian and Dutch companies were fraught with frictions. As for the relationship with Dutch companies, these frictions resulted from a lack of clarity in the terms of collaboration agreed between the governments. Early talks between the West German and Dutch governments envisaged that the Dutch would supply two-thirds of the sodium components. The guidelines of collaboration as agreed in 1968 were not clear on this point: the English version simply said that the Dutch would supply sodium components, but the German version formulated "the sodium components." The Dutch's claim to supply all sodium components triggered a considerable degree of politicking, until the Germans finally gave in.[56]

The Germans tried a similar pushing and pulling with regard to fuel fabrication, though with less success. Soon after the government memoranda were exchanged, Belgonucléaire proposed to Alkem of West Germany to form a joint company with two fabrication plants: one in Belgium, one in West Germany. Alkem did not accept this proposal, but entered into an agreement with Belgonucléaire that gave each company a share in the orders received by the other. However, Alkem handled the agreement in such a way that Belgonucléaire preferred to get out of it at the earliest possible date.[57] Frictions about fuel fabrication did not have a significant impact on the SNR-300. Nevertheless, their consequences show up in the fact that it needed prodding by the utilities to make the two fuel fabricators agree on identical design specifications for the fuel.

The settlement on the sodium components had some features that in retrospect appear unsatisfactory. As mentioned earlier, the utilities for technical and commercial reasons would have liked to see some competition between Dutch and West German subsuppliers. In their view the international collaboration reduced technical diversity too early.[58] The supply consortium had offered for the SNR-300 plant only straight-tube steam generators of Dutch design, but the utilities deemed it wise to test also helical-tube steam generators in the SNR-300 plant, and to make a choice between these two types only for the subsequent plant. Furthermore, the utilities were not happy to practically face a monopoly, as the Dutch tenders for large sodium components were up to 30 percent higher than West German tenders. The agreement between the governments was not flexible enough to deal with such a situation. It would therefore have been more satisfactory to determine in the agreement that each country should receive a share on the plant supplies according to its financial contribution, and leave open the definition of these supplies; or to specify the supplies from each country in advance and provide that each country pay for them regardless of their price.

The reluctance of industrial firms to collaborate shows clearly in the deliberations of the advisory bodies. As mentioned earlier, the advisory body recommending government support for the SNR-300 considered the possibility of constructing instead of the SNR-300 a much larger plant in collaboration with Britain or France. This option was ruled out because the consortium found a junior role in such a project not acceptable, as long as construction of the SNR-300 was not precluded by prior decisions of the three governments. It appears that the definition of acceptable terms for domestic industry is very much a function of available alternatives or, in the words of the subcommittee, of "prior policy decisions." If the government makes the funds available for an independent project, domestic industry will prefer this to a collaborative project. If the government is determined, however, to share the costs with other governments, domestic industry will put up with collaboration but will prefer a senior role to a junior role. And if prior policy decisions are taken, firms will finally find a junior role acceptable. It follows from this logic that, in state-financed development programs, the willingness of firms to collaborate with foreign industries is dependent on the prospects for continuing support from their governments.

Even with government support assured, firms may put up with a junior role if they can trade this for commercial benefits. This is illustrated by Siemens's move for a closer collaboration with the French in 1969. When the French decided to abandon their gas-cooled graphite-moderated reactors in favor of LWRs, Siemens offered the French government a share of 30 percent in Kraftwerk Union and leadership in a joint French–West German fast breeder program. This deal, proposed in November 1969 in close collaboration with the West German Science Ministry, would have given Kraftwerk

Union leadership in the French LWR market, against which the West Germans were prepared to accept a junior role in fast breeder development. Although the French were not disinclined toward a West German minority participation in their fast breeder program, they were not willing to leave their LWR market to a West German-controlled company. The talks ended in mid-1970 before the possible effects of the proposed deal on the West German–Belgian–Dutch fast breeder program could be clearly seen. Thus it is an open question, whether or not this deal would have meant cancellation of the SNR-300 plant. Nevertheless, the proposal showed that the Siemens management was prepared to give up West German independence in fast breeder development for West German domination of the French LWR market.

In general, in the 1960s there was little inclination on the part of the French and British to collaborate with the West German–Dutch–Belgian fast breeder program. Britain and France were reluctant to irradiate fuel pins for the West Germans in the DFR and Rapsodie reactors. The collaborative spirit of those years is highlighted by the fact that the British finally permitted the irradiation of some West German–Belgian fuel pins, but not of the spacer grid connecting the fuel pins in a subassembly.[59]

As mentioned earlier, improving prospects for collaboration with the British were used in 1971 as the main argument for an early construction start of the SNR-300. West German and British firms in fact announced in July 1971 their intention for a "technology and market pact for nuclear reactors" involving Kraftwerk Union (KWU) of West Germany, The Nuclear Power Group (TNPG) of Britain, Agip Nucleare of Italy, Belgonucléaire of Belgium, and Neratoom of the Netherlands. Central to the pact was an outline agreement between KWU and TNPG to bid together for commercial nuclear plants on third-country markets. Concerning FBRs, the pact included a memorandum by TNPG, Interatom, Belgonucléaire, and Neratoom, in which these companies declared their intention to collaborate in design and construction. The collaboration with the British would, according to official West German intentions, have implied the joint construction of a large demonstration plant, so there would possibly have been West German–Belgian–Dutch participation in the British CDFR. In 1972 an agreement between TNPG and Interatom for establishing a joint marketing and manufacturing company in the fast breeder sector was almost ready to be signed. However, talks ended soon after the reorganization of British nuclear industry which finally led to the replacement of TNPG and a second nuclear consortium by a new company. Thus one of the main arguments for a quick construction start for the SNR-300 had vanished before construction actually started.[60]

Whereas the British rejected in the early 1970s the German proposals for collaboration, it is interesting to see that international collaboration was later used in Britain as an argument for building the CDFR plant in the same manner as it was used for the SNR-300. Without construction of the CDFR, it was said, Britain would be excluded from the fast breeder club and would not be in a position to exchange experience with others.[61] No matter whether this argument was tactical or serious, international collaboration seems to increasingly replace international competition as a rationale for proceeding with a domestic development effort.

While the manufacturing firms in the West German-Belgian-Dutch fast breeder program were negotiating with British organizations, the utilities looked for collaboration in other directions. In May 1971, RWE Electricité de France, and ENEL (Italy) agreed to cooperate in the construction and operation of two commercial-size demonstration FBRs, one to be built in France, the other in West Germany.

Further details of the cooperation of the three utilities were fixed in an agreement signed in December 1973. The French demonstration plant Super-Phénix was to be built by a French subsidiary, in which Electricité de France holds 51 percent, ENEL 33 percent, and RWE 16 percent. This company was founded in July 1974 and was called Centrale Nucléaire Européenne à Neutrons Rapides S.A. (NERSA). The German demonstration plant SNR-2 will be owned and operated by a German firm in which RWE holds 51 percent, ENEL 33 percent, and Electricité de France 16 percent. This company was formed in October 1974 under the name Europäische Schnellbrüter-Kernkraftwerksgesellschaft mbH (ESK). RWE acted in these agreements virtually as a representative of SBK, to which it transferred its shares at a later date.[62]

On the basis of a joint declaration by the governments of France and West Germany signed in February 1976, collaboration was established also on the level of supplier firms and government laboratories. Contracts signed in July 1977 provide for pooling know-how, for cooperation of INB and Novatome in the planning and construction of FBRs, and for cooperation of French and German organizations in a coordinated R&D program (with the association of Belgian, Dutch, and Italian organizations).[63]

International collaboration is often criticized in that it makes a commitment irreversible, so without the consent of its partners a country cannot withdraw from a collaborative project. It was not until 1977 that this kind of inflexibility was officially stated in the West German–Belgian–Dutch fast breeder program. In a panel discussion published in January 1977, Hans Matthöfer, the German Minister for Research and Technology, said that a termination of the SNR-300 plant would not save money for the West Germans, as in that case they would have to repay the Belgians and the Dutch their contribution to this plant. "If I stop the work, I do not save a single deutschmark; on the contrary, I will probably have additional expenses."[64]

Obviously this statement was given without prior consultation of the Dutch and Belgian partners on their conditions for a consent to a termination of the SNR-300 project. The press reported in the same month that the West Germans pressed the Dutch to stay in the SNR-300 project.[65] Although the extent to which the Belgians and the Dutch would really resist a termination of the SNR-300 project is an open question, the inflexibility of collaboration seems to provide an easy argument for publicly justifying continuation of the project.

The collaboration with the French agreed in February 1976 will help to create more arguments of this sort. The declaration signed by the West German Ministry for Research and Technology and the French Minister for Industry and Research states that the two countries will allocate to this collaboration substantial funds of comparable amounts within the framework of available appropriations.[66] Although this clause does not bind the parliament in its appropriations to fast breeder activities, it does imply a commitment of the ministry. Once the West German government has paid its contribution to Super-Phénix, it will be possible to argue that the SNR-2 must be built in order to get this money back.

Financial Commitments

The contributions of West German firms from their own funds to the SNR-300 plant are small, both in relation to total project cost and in relation to their financial capability (see table 6-3). By the time construction started, the commitment of manufacturing firms (Interatom and Alkem) amounted to DM10.5 million. This was a contribution to the industrial R&D for the SNR-300 that was then estimated to cost DM133 million, the rest being financed by the West German government. By 1980, estimated total cost of this R&D program had increased to DM303 million, while industry had raised its financial contribution to DM24 million.[67]

As the West German supplier firms manage a major part of the construction contracts for the SNR-300 plant, it seems that the firms can finance their contribution to R&D from the usual profit margin, provided that penalties for cost-overruns do not grow large. The supplies at cost price for the SNR-300 include a 5-percent profit margin. Supplies at fixed price (plus inflation) are usually calculated with a slightly higher profit margin.[68]

A perspective of the capability of manufacturing industry to finance R&D from its own funds may be given by Siemens's annual R&D budget. In the business year 1976–1977, the Siemens concern, including Kraftwerk Union and its subsidiaries, spent about DM2,100 million on R&D, of which 92 percent was self-financed.[69] If industrial R&D for the SNR-300 plant is spread over a ten-year period, the DM24 million contribution by West

Table 6–3
Expenditure on Fast Breeder Development in West Germany, by Source of Funds

Project	Period	Total Expenditure (with foreign contributions)	West German Expenditure by Source of Funds				
			Total	Government	Manufacturing Industry	Utilities	Loans
SNR-300							
Design [a]	1966–1971	n.a.	95	95	—	—	—
Construction	1972–1985	3,713	2,599	2,151	—	148	—
R&D during constr.	1971–1984	n.a.	303	279	24	24	—
Fuel		93	53	53	—	—	—
Total		n.a.	3,050	2,878	24	148	—
SNR-2							
Design	1972–1986	50	34	—	—	34	—
R&D (preconstr.)	1973–1982	n.a.	162	146	16	—	—
Total		n.a.	196	146	16	34	—
Super-Phénix							
Construction	1976–1983	5,500	615	55	—	168	392
KNK							
I	1961–1976	173	173	164	5 [c]	4	—
II	1968–1980	180	179	179	—	—	—
Operation	1970–1980	119	119	119	—	—	—
Research laboratories	1960–1982	n.a.	1,609	1,609	—	—	—
Miscellaneous [b]	1968–1971	n.a.	11	11	0.02	—	—
Grand Total		n.a.	5,952	5,161	45	350	392
(in percent)			(100)	(86.7)	(0.8)	(5.9)	(6.6)

Sources: *Bundeshaushaltsplan 1970–1980*; Arbeitsgemeinschaft der Grossforschungseinrichtungen (AGF), *Programmbudget 1979* (Bonn: AGF, no date); interviews and correspondence with participants, November 1979, March 1980, July 1980.

Note: Actual and planned expenditure, in million DM at current prices.

[a] Includes R&D during the design phase.

[b] Carbide fuel, sodium/carbon dioxide cooling, integrated steam-generator.

[c] Not including DM10–20 million unplanned expenditure due to cost-overruns in a fixed-price contract.

German supplier firms to the SNR-300 would take 0.1 percent of Siemens's self-financed R&D. All industrial R&D associated with the SNR-300 would take 1.6 percent of Siemens's self-financed R&D.

Another comparison may be done with Siemens's R&D expenditure on first-generation nuclear reactors. Knowledgeable sources estimate Siemens's expenditure of own funds from 1956 to 1967, that is from the start of its nuclear-power activities up to its first commercial order, at around DM100 million.[70] This investment was the equivalent of 0.17 percent of the company's sales income over this period. This compares to 0.01 percent for liquid-metal fast breeders, which is a different order of magnitude.[71]

It should be stressed that the nuclear plant business was in an uncomfortable situation in the early 1970s, when the government invited the firm's financial participation in the SNR-300 project. AEG was sustaining heavy losses on its nuclear plant contracts, as tightening licensing standards, technical mishaps, and inflation engendered high costs. Because of fixed-price contracts, the company could not transfer these costs to its customers. While Siemens had much less trouble than AEG, it did not make profits on its early nuclear plant contracts. Kraftwerk Union, up to 1976 a joint subsidiary of AEG and Siemens, since 1977 fully owned by Siemens, showed losses up to 1975. This situation made it very difficult for the ministry to extract even a small financial contribution from the manufacturers. Nevertheless, comparisons with Siemens's R&D budget and its commitment to LWRs show that the limiting factor was the firm's economic assessment of the FBR rather than its capability of financing R&D.

The commitment by West German utilities from their own funds to the SNR-300 was DM84 million as estimated in 1973, or DM148 million as estimated in 1980. If DM84 million are distributed over an estimated construction period of five years, this contribution takes 0.18 percent of RWE sales as of 1973–1974.[72] Neither the higher contribution to SNR-300 as of 1980 nor RWE's contribution to Super-Phénix alter this picture, if they are distributed over the corresponding construction periods, as sales of RWE have grown considerably in the 1970s.[73] A perspective may be provided by the expenditure of RWE for the VAK and KRB plants. Both plants took 0.7 percent of RWE's sales over the construction period, about four times more than the utility's commitment to the SNR-300.

The utilities' financial contribution may have been restrained by the fact that West German laws regulating the utility industry do not explicitly state R&D as a function of the utilities. In negotiations with the government, utilities have repeatedly resisted a greater contribution to government-sponsored R&D projects on this ground.[74] This would imply that the utilities' contribution may have been restricted by legal factors. This would be consistent with the view that the limiting factor was not their financial capability. But the utilities' commitment to LWRs show that their contribution to the FBR was not limited by legal factors.

With regard to their incentives for a commitment to fast breeder development, utilities stress that the objective of a domestic development effort is not to have cheaper electricity or better security of supply but to have FBRs, should they eventually become attractive, supplied by a domestic instead of a foreign manufacturer.[75] Nevertheless, it is advantageous to the utilities to have a supplier monopoly guarded against by the existence of a number of suppliers, and to have among them a domestic supplier. Their financial commitments show that the objective of having a domestic supplier was accorded a higher priority for LWRs than for FBRs.

The Ministry and Its Clientele

Contrary to earlier decisions in the West German fast breeder program, the decision to construct the SNR-300 plant did not need an initiative, but resulted from the program's inner momentum. After the design of the SNR-300 plant had been submitted, it was natural that the plant would be constructed, unless there were pressures for a drastic policy change, such as abandoning this prototype plant in favor of a broader international collaboration or terminating the fast breeder program altogether. As the decision finally came out, it deviated little from the course of events traced out by the program's institutional momentum. The only change effected was a delay of construction start by two years.

Although some alternatives were discussed within the participating organizations, none was put forward in a manner which could have excited controversy. The utilities considered for a while construction of a 600-MWe plant with British collaboration as an alternative to the SNR-300 plant. But this idea was not formally proposed to the ministry. The utilities abandoned it after the ministry had solved the fuel-price problem to their satisfaction.

The Karlsruhe laboratory had suggested some projects complementing the SNR-300 plant, such as the FR3 reactor and the gas-cooled fast breeder. These were always proposed with the understanding that the SNR-300 should have first priority.

There were no political incentives or rewards for altering the course of things. Compared to the decision in 1969 to terminate work on the steam-cooled fast breeder, a decision avoiding the construction of the SNR-300 would have been much more difficult. While in the decision on the steam-cooled fast breeder at least part of the ministry's clientele was pressing for a reversal of policy, there was wide agreement that the SNR-300 plant should be constructed independently from collaboration with the British or French, and that construction should start expeditiously. In contrast to the steam-cooled fast breeder, the sodium-cooled fast breeder was not an isolated national development. As Britain and France were already constructing large liquid-metal fast breeder prototypes and the United States was planning such a plant, a cancellation of the SNR-300 would have excited world-wide discussions. Without compelling evidence that the liquid-metal fast breeder would be a failure, such a discussion was too adventuresome for ministry officials to engage in. The premium was clearly in not trying to be better than their foreign colleagues and in staying on the international bandwagon.

While the steam-cooled fast breeder was one of several options within the fast breeder program, the SNR-300 now formed the very center of the program. The central question was whether or not to keep the industrial teams alive. If the teams were to be kept in good shape, there had to be a construction project within a few years. Therefore, among ministry officials the attitude prevailed that the SNR-300 project should be carried on as long as difficulties did not prove insuperable. In the negotiations, the ministry thus removed a number of obstacles over which the project would otherwise have stumbled.

This attitude was in sharp contrast to the ministry's hard stance on other projects in the fast breeder program. The FR3 research reactor, for example, was aborted by the ministry before its advisory bodies formulated a recommendation. When the Working Group of the German Atom Commission on 6 March 1971 discussed the FR3 proposal, representatives of the ministry told the committee that the estimated cost of DM400 million could not be managed by West Germany and the Benelux countries in addition to the SNR-300, so the reactor could be realized only through broader international collaboration.[76] This assured a quiet death for this reactor, as Britain and France with their fast breeder prototype plants then in operation or under construction commanded sufficient facilities for test irradiation in fast neutron flux. Their disinterest in a larger research reactor was predictable. Another example is the gas-cooled reactor. When the Working Group in 1971 discussed a DM20 million R&D program for this technology over three years, ministry officials placed beforehand a clear limit on the funds available. They proposed that the planned R&D program was to be financed from the yearly budget of the government laboratories. As the government was unable to allocate additional funds, this finance would have to be made available through cuts in other activities. Furthermore, they stated that for the next two years the ministry would not allocate any government money to industrial firms for work on the gas-cooled fast breeder.[77]

As to the SNR-300, ministry officials did not really expect from the consultations of the advisory bodies an answer on whether to construct or not. Though ministry officials participated in the subcommittee's final written report, it was not their intention to secure a positive recommendation but to have the foreseeable positive recommendation justified by as detailed arguments as possible. It seems, however, that their demands for a detailed justification were not exorbitant. They were satisfied with an analysis which, for the technology and economics of FBRs, was based only on evidence supplied by the reactor manufacturer. Although speakers of the subcommittee found it necessary to have this information reviewed by the utilities, the officials accepted a recommendation of the Working Group based on this limited evidence.

Nevertheless, the justification proved better than that required by higher instances of policymaking. Neither the higher levels of the ministry, nor the cabinet, nor the Bundestag challenged the SNR-300 decision. This decision would have made its way through the formal policymaking process with a less detailed justification than that actually provided.

The drastic increases in the costs of the SNR-300 plant in 1971–1972 somewhat embittered the ministry's relations with the Karlsruhe laboratory and the supplier company. As the earlier cost projections by the Karlsruhe laboratory and by Interatom had proved so markedly wrong, ministry officials had become increasingly critical of any advice from those sources. Politicians and journalists wanted to know why the cost figures of the SNR-300 differed so much from the earlier estimates. Explaining this was not a pleasant job for ministry officials.[78]

In both the Karlsruhe laboratory and supplier industry the period 1970–1972 brought also some internal tensions. In Karlsruhe, the need to revise so drastically the earlier cost estimates added to the internal opposition to the project management. As a result of pressures from inside and from the ministry, in 1970 the project management was reorganized structurally, with a reduction in the project manager's influence. Finally, these pressures led to personnel changes in the project management. The negotiations on the SNR-300 also affected the relations between Interatom and Kraftwerk Union, as the latter had to give guarantees for its subsidiary. Kraftwerk Union decided to make some changes in the management of Interatom, when the company entered the SNR-300 contract.

In the negotiations on construction of the SNR-300 plant and on the preparations for the SNR-2 plant, the utilities, as the future users of the plants, assumed an active role for the first time in the West German fast breeder program. Their participation in the decision-making process resulted in drastic alterations in the SNR-300 design, adapting it to user needs, reducing operating costs, and increasing the technical and economic information to be gained from construction and operation of the plant. These alterations have notably improved the SNR-300 design. This observation is consistent with the results of empirical studies on industrial innovation, which found that attention to user needs discriminated more between failures and successes than any other factor investigated.[79]

The utilities' participation in the negotiations on the SNR-300 contracts had also positive effects on the negotiated price. The price offer of the consortium was reviewed by both the ministry and the utilities. Although the ministry's review did make a contribution to the price reduction achieved in the subsequent negotiations, the major part of this reduction, as far as it was real rather than cosmetic, was due to the commercial expertise and bargaining skill of the utilities.[80]

The utilities' participation also contributed to a deeper assessment of technical and commercial risks. This can be seen in the utilities' insistence on a design sufficiently detailed to allow an assessment of the price worthiness of the consortium's offer and in their extraction of technical guarantees from the member companies of the consortium, including Kraftwerk Union as the parent company of Interatom.

Although the utilities' influence in the negotiations for the SNR-300 was important, it was exercised within clear limits. The utilities resolutely used their influence only on those questions which were critical for keeping the technical and commercial risks of the SNR-300 plant at an acceptable level. They were prepared to withdraw from the project if certain preconditions in this regard were not met. These points had to be cleared, even if the necessary negotiations resulted in a delay of the construction start of the SNR-300 plant.

Their attitude was different, however, with regard to their demands for design alterations resulting from user needs and with regard to their wishes concerning industrial policy. Here they were prepared to settle for a compromise which could be achieved without postponing construction. Thus the utilities dropped their demands for a second supplier of steam generators or for a higher rating of the steam generators when it became clear that these could not be realized without substantial delay in the construction schedule.

There are several reasons why the utilities became so influential: they were to contribute the largest amount from their own funds, and their advice was not vitiated by previous erroneous cost estimates. But the most important reason was that industrial policy in West Germany obliged the ministry to seek the utilities' collaboration and that the utilities were relatively independent of the ministry, unlike the Karlsruhe laboratory and Interatom. This independence is illustrated by the fact that the utilities went much farther in communicating to the public their skeptical economic assessment of the FBR than the ministry liked to see. At the 1973 status report a utility executive outlined some economic uncertainties in the fast breeder fuel cycle. He implied that under prevailing economic conditions plutonium recycle was perhaps not feasible for LWRs. Ministry officials reacted with anger and reproach.[81]

Nearly all interviewees from the government, the suppliers, and the utilities agreed that the utilities should have assumed an influential role earlier.[82] Managers in the supplier industry stressed that they sought the utilities' opinion on design questions long before the design was submitted at the end of 1969. Ministry officials said that they had urged the utilities to participate more closely in the design activities for the SNR-300.

A clear answer as to why the utilities assumed their role so late did not emerge in the research for this study. Whereas some factors did contribute to this situation, their relative importance cannot be assessed. One such factor was the long delay in the decision about which of the West German utilities would participate in the project and function as the owner/operator of the plant. The ministry would have liked a large number of utilities to participate in the project, as it wanted to spread the know-how generated in this project over as many customers as possible. However, the utilities were reluctant to participate in the project. Many companies seemed glad to let RWE finally take the lead. RWE, as the biggest utility, was the traditional market leader anyway. Its executives were in the Fast Breeder Project Committee as well as in the Working Group of the German Atom Commission.

RWE itself, once it had decided to engage in the project, seems to have preferred a dominant role in order to facilitate the negotiations on construction and operation of the SNR-300 plant. With a 70-percent stake in SBK, RWE led these negotiations on the utility side. Substantial collaboration by other West German utilities would have necessitated a lot of coordination. It might also have generated friction and possibly have weakened the bargaining position of the utilities versus the supplier. After the contracts for the SNR-300 were settled, RWE issued an invitation to other West German utilities to participate in the contracts. However, it received no positive responses. The hesitation by the West German utilities to participate in the project does not indicate great enthusiasm about fast breeder development.

Another factor working against a more intensive participation by the utilities in the period before 1970 was their limited staff. Because of the long-standing tradition of turnkey contracts, West German utilities have had a relatively small staff for managing their activities in plant construction. During the 1960s this also applied to nuclear plants. They were hesitant to occupy their limited staff with details of fast breeder development or to employ additional staff for this purpose. This would indicate a reserved attitude of the utilities toward fast breeder development.

The organizational arrangement for the SNR-2 design shows that lessons from the SNR-300 experience have been learned. The design for the SNR-2 has been contracted by the utilities to Interatom, later INB. The costs are shared by government and the utilities. Apart from the project definition phase, the design contracts are paid for by the utilities out of their own funds, while the government pays for the supporting R&D program. The utilities are therefore involved in a formal way, giving them the power to determine design specifications for the SNR-2. As the design contract is handled along the usual lines of business, the expertise of their staff is brought to bear upon the design specifications. In retrospect, it can be said that such an organizational arrangement would also have been appropriate for the SNR-300 design.

Notes

1. The name of the ministry responsible for the fast breeder program changed several times in the period covered by this chapter: the exact names were Ministry for Scientific Research (up to Ocotber 1969), Ministry for Education and Science (up to December 1972), and since then Ministry for Research and Technology.

2. For published accounts see U. Däunert and W.J. Schmidt-Küster, "Das Projekt SNR—staatliche Förderung und internationale Zusammenarbeit," *Atomwirtschaft* 17 (1972):363-366; K. Traube, "Der SNR-300 und die internationale Situation der Schnellbrüterentwicklung," *Atomwirtschaft* 18 (1973):411-414.

3. Interviews with participants, December 1974, February 1975, March 1977, October 1977, November 1977. See also "Deutsch-niederländisch-belgische Zusammenarbeit bei der Entwicklung schneller Brutreaktoren," *Pressedienst des Bundesministeriums für wissenschaftliche Forschung,* no. 13/14 (26 July 1967):119.

4. "Europäische Zusammenarbeit der Industrie bei der Schnellbrüter-Entwicklung," *Pressedienst des Bundesministeriums für wissenschaftliche Forschung,* no. 1 (24 January 1968):6; R. Ernst and E.A. Guthmann, "Europäische Zusammenarbeit bei der Entwicklung des natriumgekühlten Schnellbrüters," *Atomwirtschaft* 14 (1969):448-450.

5. Foratom, *The Nuclear Power Industry in Europe* (Bonn: Deutsches Atomforum, 1972), pp. 102-108.

6. Ibid.; Centre de Recherche et d'Information Socio-politiques (CRISP), "Le Secteur Nucléaire en Belgique: Développement et Structures Actuelles," *Courier Hebdomadaire,* no. 718-719 (Brussels: CRISP, 23 April 1976); interview with a participant, November 1977.

7. Foratom, op. cit., note 5, pp. 233-235; *Interatom: A Portrait of the Company,* prospectus issued by the company, December 1974; "Neue Beteiligungsverhältnisse bei INTERATOM," *Atomwirtschaft* 14 (1969): 278-279; interviews with participants, January 1975, February 1975, December 1975, September 1976.

8. E.A. Guthmann et al., "The Fast Breeder Development Program of the Konsortium SNR," Paper P/369, *Fourth United Nations Conference on the Peaceful Uses of Atomic Energy, Geneva, 6-11 September 1971,* proceedings, p. 89; Ernst and Guthmann, op. cit., note 4, p. 449.

9. For the LWR demonstration plants see chapter 2.

10. For published information on the organizational framework of the SNR-300 see G.H. Scheuten, "Organisation und wirtschaftliche Bedeutung der Brüterentwicklung," *Atomwirtschaft* 17 (1972):366–368; A Brandstetter and A.W. Eitz, "The Tripartite Fast Breeder Programme: A Utility/ Industry View," *Nuclear Engineering International* (July 1976):40–43.

11. Kernforschungszentrum Karlsruhe, Projekt Schneller Brüter, "Ausführliche Erläuterungen zu den Anträgen zur Bereitstellung der Mittel für die Erstellung der Unterlagen zum Bau der Prototypen des Schnellen Brüters," unpublished report, October 1965, pp. 38–43.

12. Interviews with participants, February 1975, February 1976.

13. Interviews with participants, December 1974, January 1975, February 1975, December 1975.

14. Foratom, *The Nuclear Power Industry in Europe,* (Bonn: Deutsches Atomforum, 1978), pp. 391–392.

15. Interviews with participants, March 1977, October 1977.

16. W. Häfele, "Die Entwicklung Schneller Brutreaktoren und der SNR 300," *Atomwirtschaft* 16 (1971):247–250; W. Häfele and G. Kessler, "SNR: The German-Benelux Fast Breeder," *Nuclear News* (March 1972):48–53.

17. A. Brandstetter and H. Hübel, "Sicherheitskonzept und zugehörige Konstruktionsmerkmale des SNR-300," *Atomwirtschaft* 17 (1972): 371–374; J.M. Morelle, K.-W. Stöhr, and J. Vogel, "The Kalkar Station: Design and Safety Aspects," *Nuclear Engineering International* (July 1976):43–49. In addition to published information, the following section draws on interviews with participants, December 1974, January 1975, February 1975, March 1976.

18. In connection with licensing requirements for the suspension of the reactor vessel, it was later decided to raise the specification for the reactor vessel to 370 MWsec. It proved feasible to accommodate this power excursion without changing the design of the vessel. See K. Traube, "The Benelux-German Fast Reactor Programme," *Journal of the British Nuclear Energy Society* 15 (1976):215–224.

19. This section is based on interviews with participants, December 1974, February 1975, December 1975.

20. The initial design is described in some West German contributions to *Sodium-Cooled Fast Reactor Engineering: Proceedings of a Symposium, Monaco, 23–27 March 1970,* edited by the International Atomic Energy Agency (Vienna: International Atomic Energy Agency, 1970).

21. "SNR-300:niedrigere Brutrate zur Kostenersparnis," *Atomwirtschaft* 20 (1975):261–262.

22. W. Häfele and K. Wirtz, "Ansätze zur Entwicklung eines schnellen Brutreaktors," *Atomwirtschaft* 7 (1962):557–559.

23. Interviews with participants, March 1975.

24. D. Smidt et al., "Referenzstudie für den 1 000 MWe Natriumgekühlten Schnellen Brutreaktor (Na 1)," Report KFK-299 (Karlsruhe: Gesellschaft für Kernforschung, December 1964); K. Benndorf et al., "Variation einiger wichtiger Reaktorparameter beim natriumgekühlten 1 000 MWe Schnellen Brüter zur Untersuchung der Brennstoffkosten und des Brennstoffbedarfs," Report KFK-568 (Karlsruhe: Gesellschaft für Kernforschung, July 1967).

25. This section is based on interviews with participants, December 1974, January 1975, February 1975, December 1975, May 1977, October 1978, June 1979, March 1980.

26. "GE Tells ESADA Breeder Demo Would Cost $400–500 Million," *Nucleonics Week* (23 April 1970); "Das Ende der Natriumbrüter-Illusionen," *Frankfurter Allgemeine Zeitung* (6 May 1970).

27. G. Vendryes and R. Carle, "French Fast Breeders," *Proceedings of the American Power Conference* 34 (1972):83–88. These authors give a figure of $114 million (in 1972 dollars); at official exchange rates this was DM364 million. As the authors mention, a similar estimate was quoted by Vendryes in June 1970.

28. See also Däunert and Schmidt-Küster, op. cit., note 2, p. 365.

29. K. Gast and E. Schlechtendahl, "Schneller Natriumgekühlter Reaktor Na 2," Report KFK-660 (Karlsruhe: Gesellschaft für Kernforschung, October 1967), p. 8-1.

30. This section draws on interviews with participants, February 1975.

31. Working Group III/1, on Reactors, of the German Atom Commission (henceforth Working Group), minutes of 64th session, 6 March 1970, pp. 12–14 (unpublished).

32. Working Group, minutes of 66th session, 27 January 1971, p. 10 (unpublished).

33. "Bericht des Ausschusses 'Förderungswürdigkeit des SNR-300' an den Arbeitskreis III/1 'Reaktoren' der Deutschen Atomkommission," (unpublished).

34. Working Group, minutes of 67th session, 28 October 1971, pp. 5–10 (unpublished).

35. "BMBW: Abschlusssitzung der Deutschen Atomkommission," *Pressedienst des Bundesministeriums für Bildung und Wissenschaft,* no. 41/71 (20 October 1971).

36. Interview with a participant, October 1976.

37. Deutscher Bundestag, "Öffentliche Informationssitzung des Ausschusses für Bildung und Wissenschaft am 17. Dezember 1970, Anhörung von Sachverständigen zu dem Thema 'Wachstumsorientierte Technologien und staatliche Forschungspolitik' betr. den Bereich Kernenergie," see especially pp. 5–16, 71, 117.

38. Deutscher Bundestag, Kurzprotokoll 5. Sitzung der Arbeitsgruppe

"Neue Technologien" des Ausschusses für Bildung und Wissenschaft, 12. März 1971, p. 5/6 (unpublished). The designation of the subcommittee and the number of sessions is not used in a consistent manner in the minutes. As membership and chairmanship in the fast breeder subcommittee coincided with that of two other subcommittees, topics of the three subcommittees were sometimes dealt with at one session.

39. Deutscher Bundestag, Ausschuss für Bildung und Wissenschaft, minutes of 38th session, 29 April 1971, pp. 30–33 (unpublished).

40. Deutscher Bundestag, Ausschuss für Bildung und Wissenschaft, Arbeitsgruppe "Schneller Brüter", minutes of 8th session, 9 June 1972 (unpublished), and interview with a participant, May 1977.

41. Däunert and Schmidt-Küster, op. cit., note 2.

42. Ibid., p. 365.

43. Traube, op. cit., note 18, p. 216.

44. H. Böhm et al., "Irradiation Behavior of Fast Reactor Fuel Pins and their Components," Paper P/392, *Fourth United Nations Conference on the Peaceful Uses of Atomic Energy, Geneva, 6–11 September 1971,* proceedings, p. 24.

45. K. Traube, "Stand der industriellen Arbeiten zum SNR-300," in *Statusbericht 1974,* edited by Kernforschungszentrum Karlsruhe, Projekt Schneller Brüter, Report KFK-2003 (Karlsruhe: Gesellschaft für Kernforschung, March 1974), pp. 21–34.

46. J.P. van Dievoet, "The Fuel for SNR-300," *Nuclear Engineering International* (July 1976):54–55.

47. Häfele, op. cit., note 16; P. Engelmann, "Die Arbeiten des Karlsruher Projektes Schneller Brüter," *Atomwirtschaft* 16 (1971):251–255.

48. W. Häfele, "Die Entwicklung Schneller Brutreaktoren und der SNR 300," *Atomwirtschaft* 16 (1971):247–250. Reprinted with permission.

49. Interviews with participants, February 1975, October 1976.

50. Interviews with participants, February 1975, September 1976, October 1976, May 1977.

51. "Wo steht die deutsche Schnellbrüterentwicklung? Die öffentliche Diskussion," *Atomwirtschaft* 16 (1971):195–208.

52. Working Group, minutes of 67th session, 18 October 1971, p. 4 (unpublished).

53. Traube, op. cit., note 45, p. 29; idem, op. cit., note 18.

54. Interviews with participants, December 1975.

55. Interviews with participants, January 1975, September 1976.

56. Interviews with participants, December 1975, June 1977, October 1977, November 1977.

57. Interviews with participants, October 1977.

58. Interview with a participant, January 1975.

59. Interview with a participant, September 1976.

60. "Four-Nation Nuclear Pact to Boost Sales and Knowledge," *The Guardian* (15 July 1971). See also "Europäische Atomreaktor-Hersteller wollen Wettbewerbsfähigkeit verbessern," *Die Welt* (15 July 1971); "Kooperation im Reaktorbau," *Handelsblatt* (15 July 1971); "UK and Germany Set to Sign Nuclear Reactor Agreement," *The Times* (London: 15 November 1971); Traube, op. cit., note 18, p. 221.

61. "Reactor Decision Left to Government," *Financial Times* (23 September 1976).

62. "Accord Franco-Allemand pour la Construction de Surrégénérateurs," *Le Monde* (10 May 1971); "L'Italie se joint à la France et á l'Allemagne pour Construire Deux Réacteurs Surrégénérateurs de 1 000 Mégawatts," *Le Monde* (29 June 1971); "Europäisches Brüterreaktorprojekt," *Neue Zürcher Zeitung* (10 February 1974); "Brütergesellschaft NERSA gegründet," *Atomwirtschaft* 19 (1974):381; "Basis für zweiten Schnellbrüter," *Süddeutsche Zeitung* (16 October 1974).

63. A. Ertaud and J. Naigeon, "How Europe Is Cooperating on FBRs," *Nuclear Engineering International* 24 (January 1979):15–16, 21–22; G. Vendryes and H.H. Hennies, "Französische Brüterentwicklung und deutsch-französische Zusammenarbeit," *Atomwirtschaft* 23 (October 1978):448–451.

64. "Unter der Wolke des Atoms," *Die Zeit* (7 January 1977):9–13.

65. "Fast Reactor Project Gets Continued Support," *Nuclear News* (January 1977):87.

66. Deutscher Bundestag, Ausschuss für Forschung und Technologie, minutes of 44th session, 18 February 1976, appendix 1.

67. *Bundeshaushaltsplan 1973 and 1980*, Comments on Kap. 3005 Tit. 89310.

68. Interviews with participants, December 1975, June 1977.

69. Siemens, *Annual Report 1976/77*, p. 32.

70. Interview with a participant, September 1976.

71. Assuming an expenditure of DM24 million over ten years, and relating this expenditure to the 1976–1977 sales of Siemens (DM25,198 million).

72. RWE's sales in 1973–1974 were DM9,291 million; see the company's annual report.

73. The contributions of RWE to SNR-300, Super-Phénix, and SNR-2 as listed in table 6–3 add up to DM350 million. Distributed over fourteen years this takes 0.16 percent of the utility's sales as of 1978–1979 (DM15,669 million).

74. Interview with a participant, May 1977.

75. Interviews with participants, December 1975.

76. Working Group, minutes of 64th session, 6 March 1970, pp. 11–12 (unpublished); see also P. Engelmann and U. Däunert, "Present State of

Development of the Fast Breeder Project in the Federal Republic of Germany and in the Benelux Countries and Progress Achieved during 1971 and early 1972," in International Atomic Energy Agency, International Working Group on Fast Reactors, Fifth Annual Meeting, Vienna, 19-21 April 1972, Summary Report, part 2, pp. 7-29.

   77. Working Group, minutes of 66th session, 27 January 1971, p. 9 (unpublished).

   78. See, for instance, Deutscher Bundestag, Auschuss für Bildung und Wissenschaft, Arbeitsgruppe "Schneller Brüter," minutes of session 4a, 26 May 1970 (unpublished).

   79. R. Rothwell et al., "SAPPHO Updated—Project SAPPHO Phase II," *Research Policy* 3 (1974):258-291.

   80. Interview with a participant, September 1976.

   81. Interview with a participant, March 1980.

   82. Interviews with participants, December 1974, January 1975, February 1975, March 1975, February 1976, October 1976.

7 A Strategy for Future Policymaking

The economic assessment of fast breeder reactors (FBRs) has a dismal record, not only in West Germany but in most countries where the technology is being developed. In spite of dramatic increases in the price of uranium and other primary fuels, the schedule for market introduction has been delayed by decades. A development effort of twenty or thirty years has failed to bring us nearer to the commerical use of fast breeders. Advocates of the fast breeder presently see the industrial use of this technology as distant in the future as they did twenty years ago.[1] Given the dismal record of past economic assessments, a central question for future policymaking is how to produce a realistic economic evaluation.

Money and the Quality of Advice

If future decisions are to be based on sound economic judgment, three things are required: first, that the organizations involved are able and willing to make realistic economic assessments; second, that they are willing to communicate these assessments to policymakers; third, that policymakers have incentives to take these assessments into account.

As the experience in the West German fast breeder program shows, the main source of expertise are organizations with experience in the construction and operation of nuclear power plants: reactor vendors and utilities. Government laboratories and agencies are not well equipped for economic assessments because of their inexperience in commercial business. Firms without commercial experience generally are not in a better position than government laboratories.

Governmental actors have achieved a good deal of learning in the course of the fast breeder program. Ministry officials and scientists in government laboratories have become more sensitive to the uncertainties involved in the economic assessment of a new technology. Nevertheless, throughout the decisions analyzed in this study, governmental actors have been more optimistic than actors in supplier or utility industries. Even if actors on the government side finally acquire a realistic attitude toward economic and technical uncertainties, they will still depend on manufacturing firms and utilities for basic economic and technical information. Published technical literature is no substitute for firsthand knowledge.

The capability of doing something does not necessarily imply a willingness to do it. Human actors, individuals and organizations, often subject information that is favorable to their interests to less scrutiny than that which is unfavorable. This is known as bias. The optimistic bias in R&D program evaluation is a general phenomenon. Research laboratories, government or industrial, have little incentive to scrutinize information by which they can easily justify their activities to sponsors. Their willingness for a thorough and realistic assessment of a project depends on the management controls, the incentives, and the rewards imposed by the sponsors. In the age of corporate capitalism, the possibility of losing money still provides an incentive for firms to establish such management controls.

If realistic assessments are made, they are not necessarily communicated to policymakers. Only in the first decision in the West German fast breeder program did representatives of industry in the advisory committees advance their skepticism without being pressed to do so. This was in the context of a general dispute on the division of responsibilities between government laboratories and industry. In later decisions, they showed their skepticism only when asked to contribute some of their own funds. In the 1966 decision to start design on two 300-MWe prototypes, for example, industrial members in the advisory committee did not criticize the unrealistic Karlsruhe estimates of electricity costs. But when the ministry asked them to put up some funds of their own, they averred that an economic use of FBRs was not foreseeable. In the discussions about the termination of the steam-cooled FBR, industry stopped proposing its continuation only after the ministry demanded a financial contribution. A good deal of the realism that was introduced into the economic assessment of FBRs during the negotiations on the SNR-300 resulted from the fact the utilities were to finance a part of the SNR-300 from their own funds.

As long as only government money is at stake, the behavior of industrial actors with regard to the economic assessment of a program follows the principle of conflict avoidance or nonaggression. It is natural that firms depending for their survival on a government-financed R&D program have little interest in being critical. But even firms without this strong kind of self-interest have little incentive to scrutinize closely the economic prospects of a technology. There is little compensation for the unpleasantness involved in criticizing another's economic assessment. By contrast, incentives for nonaggression are manifold. Once a government agency is committed to a program, firms do not wish to be excluded by criticizing the government's economic justification. They want to be part of it in order to know what is going on and to get a share of the business involved. Open criticism would be regarded by government agencies and other firms as an unfriendly act, and a firm representative in an advisory committee has little incentive to stir up tensions between his firm and these organizations.[2]

The findings of this study suggest that firms act differently when asked to put up their own funds. An interviewee in the government sector summarized his experience by saying: "You get the virtues of industry only where industry's own money is at stake, otherwise you get only its vices."

Finally, even if skeptical assessments are made known to policymakers, they are not necessarily taken into account in policy decisions. For example, the politics of a fair return from West German contributions to Euratom made the ministry ignore the earlier recommendations of its advisory bodies. In the 1966 decision the ministry accepted the Karlsruhe estimates of electricity costs, though they contradicted industry's justification for not committing its own funds.

Decentralized Framework

These observations lend support to a decentralized framework for policy-making as proposed by Eads and Nelson (see chapter 1). According to this view, government support would be given only to exploratory R&D. The construction of demonstration plants would be made dependent on the participating firms' willingness to finance them from their own funds, while government support for these plants would be restricted to loans and to sharing operating risks. Experience in the West German fast breeder program suggests that such a decentralized framework would produce more thorough and more realistic economic assessments than in the past. It would also assure that these assessments are taken into account when decisions are made.

This decision strategy could be applied to the West German fast breeder program by saying that construction of the SNR-2 should be made dependent on the participating firms' willingness to finance design, construction, and operation from their own funds. The utilities would in any case be requested to contribute the equivalent of an LWR plant. The excess construction costs and the costs of industrial R&D and design would be shared by the supplier firms and the utilities. Government support would be confined to risk sharing in unforeseeable operation losses and background R&D performed by government laboratories.

Would such a strategy exceed the financial capabilities of the firms involved? The West German share of preconstruction industrial work for SNR-2, including design and R&D, is estimated by the supplier at DM410 million to DM430 million over a period of ten years.[3] The yearly amount is less than 2 percent of Siemens's self-financed R&D in 1979.[4] It corresponds to 0.15 percent of Siemens's sales income in that year, a fraction which approximates Siemens's self-financed expenditure on nuclear reactors prior to the company's first commercial plant order in 1967.[5] If the noncommercial costs of SNR-2 are estimated to be 50 percent of the costs of an LWR plant and spread over a ten-year period, they would take 0.88 percent of RWE's sales as of 1978–1979.[6] This is approximately the proportion RWE spent on the VAK experimental boiling-water reactor (BWR) from 1958 to 1961.[7] These figures suggest that the commercialization of fast breeders would not be beyond the capabilities of the firms involved provided this technology lives up to the promises made by its proponents.

Commercial deployment of FBRs presupposes political decisions about the acceptability of their safety and environmental risks, the social implications of large-scale use of plutonium, and the potential proliferation hazards associated with this technology. Such political questions would, in addition to technical and commercial uncertainties, create a specific risk to the companies' R&D investment. If the state wants to keep open the option to forego the commercial use of FBRs for political reasons, it would make economic sense to reduce the firms' investment risk by guaranteeing to refund a part of their R&D investment in the event of such a decision.

Would the proposed decision strategy sufficiently take into account the economic and social benefits that are claimed for the FBR? The potential benefits of a fast breeder program may be divided into two categories. First, the microeconomic benefits of having lower electricity costs than with other technologies for electricity generation. Second, the macroeconomic and political benefits derived from the greater security of supply that FBRs may be able to provide, or from the capability of domestic industry to export FBRs.

Electricity Costs

The liquid-metal fast breeder reactor (LMFBR) has advantages over the LWR in that it is virtually independent of the supply of uranium and does not need enrichment of uranium. These advantages, however, do not necessarily make the LMFBR competitive with the LWR, as they may be offset by disadvantages with regard to other cost components such as capital costs or the costs of fuel fabrication, reprocessing, and transportation. Proponents of the LMFBR usually argue that fuel-cycle costs of LMFBR plants will be lower than those of LWR plants, so the LMFBR can be allowed somewhat higher capital costs and still be competitive with the LWR.

For the present, the economics of future large LMFBR plants are a matter of estimates or guesses. However, these have some basis in the cost characteristics of the demonstration plants now under construction or in operation. The economic data available suggest that there are large uncertainties. Present official expectations for the LMFBR to become commercially competitive around the turn of the century appear too optimistic. It is not possible to say with confidence when benefits in terms of lower electricity costs will show.[8]

Capital Costs

The most important component in the costs of electricity from nuclear plants are the investment costs. In present LWRs they account for about two-thirds of total electricity costs. The SNR-300 costs about six times more per kilowatt of net electric output than a commercial LWR.[9] This is based for the SNR-300 plant on a cost estimate of June 1979. Costs have further increased since then and are likely to continue to do so until completion of the plant.

The price of the SNR-300 does not directly reflect the investment costs of larger LMFBR plants, as scaling up will bring down the costs per kilowatt. Although the cost savings to be achieved through scaling up cannot be foreseen accurately, informed guesses can be done through detailed design studies on the basis of LMFBR plants in operation or under construction. Some guidance is also provided by experience or cost studies in LWR plants. A few figures on the effects of scaling up are reported in table 7–1 for LMFBRs and LWRs. If we assume a scaling exponent between 0.4 and 0.6, a 1,300–1,500-MWe fast breeder would cost 40 to 60 percent less per kilowatt than a 300-MWe plant. The cost of SNR-300 would thus suggest that a first-of-its-kind plant in the 1,300 to 1,500 MWe range would cost between two and four times more than a commercial LWR.

The firms constructing the SNR-300 argue that this plant may not be representative for future large LMFBR plants, saying that the costs include a good deal of R&D work that will not have to be repeated for future plants; that a good deal of the SNR-300 costs resulted from design changes imposed by the licensing authorities that had to be incorporated into a largely fixed plant concept; that due to unexpected construction delays caused by the licensing procedure the suppliers' engineering capacity was underutilized, since it could not be deployed for other projects in the meantime. Finally, proponents of the fast breeder stress that series production and increasing competition among suppliers of components will reduce future costs.[10] Nevertheless, it is highly probable that most of these factors that have increased the cost of the SNR-300 plant will also be effective, though perhaps to a less extent, for the next few larger plants. These factors will become insignificant only if a series of plants of nearly identical design are constructed, that is, if each year construction of a new plant can start.

The hopes of the FBR community now rest on the French Super-

Table 7-1
Economies of Scale in Nuclear Power Plants

Type	Source	Cost Basis (year)[a]	Kind of Data	Scope of Costs	Scaling up from/to (MWe)	Reduction of Costs per kWe (percent)	Scaling Exponent[b]
LMFBR	Olds	1973	Estimate	Direct	300/1,200	45	0.57
				Indirect incl.	300/1,200	47	0.54
LMFBR	Traube	1973	Estimate	Direct	300/1,000	50	0.42
LMFBR	Guthmann		Estimate	Direct	300/2,000	55	0.58
LWR	Kraftwerk Union	1971	Price list	Direct	300/1,000	42	0.54
PWR	KWO/Biblis A	1964-1969	Actual	Indirect incl.	328/1,145	38	0.62

Sources: F.C. Olds, "The Fast Breeder: Schedule Lengthens, Cost Escalates," *Power Engineering* (June 1972): 33–35; K. Traube, "Der SNR-300 und die internationale Situation der Schnellbrüterentwicklung," *Atomwirtschaft* 18 (1973):411–414; E. Guthmann in "Streit um eine neue Technologie. Schneller Brüter: Millionen verplant?" *Bild der Wissenschaft* (1974):100–113; B. Bergmann and H. Krämer, "Technischer und wirtschaftlicher Stand sowie Aussichten der Kernenergie in der Kraftwirtschaft der BRD, 1. Teil, Fortschreibung mit Stand Oktober 1971," Report Jül-827-HT, Kernforschungsanlage Jülich, February 1972, pp. 108–109; interview with a participant, February 1975.

a Actual costs are deflated to a common price basis by the price index for capital goods.

b The scaling exponent *n* is defined by the formula $\dfrac{\text{cost of plant A}}{\text{cost of plant B}} = \left[\dfrac{\text{size of plant A (kWe)}}{\text{size of plant B (kWe)}} \right]^{n}$.

Phénix. It is reported to cost about twice as much as a French LWR plant of the same size. The hope is that series production will help reduce FBR costs to a level compatible with economic operation.[11] However, experience with LWRs suggests that it is not certain that cost savings through learning and series production will be achieved. LWR costs have increased dramatically over the last decade. Effects of learning have been canceled out by other factors such as changes in licensing requirements, increasing quality assurance and control, and schedule slippages.[12] Series production was not experienced by the LWR in the first decade after its commercial introduction. At present, increasing efforts for standardization are being made, but in all countries except France many obstacles still have to be overcome.[13] Suppliers, except General Electric and Westinghouse in the United States and Framatome in France, find it difficult to get orders in sufficient quantity to realize the benefits of series production.[14] Finally, if nuclear power plants do achieve cost savings through learning and series production, the LWR will have a fair potential for this and will provide a moving cost target for the FBR.

Therefore, capital costs of future LMFBRs are highly uncertain. Engineering estimates in recent years that take account of learning and series production give capital-cost ratios relative to LWRs between 1.0 and 1.2.[15] A more realistic study by the Rand Corporation acknowledges that a ratio of 1.2 could be reached, assuming a favorable regulatory and political environment for nuclear power. With an unfavorable environment, the cost ratio might reach 1.9, and unforeseen technical and regulatory requirements might raise it still higher.[16]

Licensing requirements will be a critical determinant for economies of scale in future LMFBRs. As the effects of the hypothetical core-disruptive accident increase with the size of the reactor, future licensing requirements relating to this accident will be a particular source of uncertainty. Present safety philosophy and assumptions on the release of mechanical work in the core-disruptive accident limit scaling up at ratings between 1,300 and 1,500 MWe. Larger unit sizes will be feasible only if assumptions on the mechanical efficiency of the core-disruptive accident prove too high and can be reduced, or if present safety philosophy is changed.

If the upward trend of nuclear plant costs continues, this alone may make it more difficult for the LMFBR to compete against the LWR. In the mid-1960s, when large LWRs were estimated to cost $120 per kWe, a 25-percent capital-cost penalty meant an electricity-cost penalty of 0.7 mills/kWh. Today, with costs of LWRs reaching $1,100 per kWe, the same capital-cost ratio of 1.25 means an electricity-cost penalty of 6 mills/kWh. What are the chances for the LMFBR to achieve lower fuel-cycle costs than the LWR in order to make up for its higher capital costs?

Fuel-Cycle Costs

The perspective that the SNR-300 throws on fuel-cycle costs is as uncertain as that on capital cost.[18] Estimated fuel-cycle costs for the SNR-300 are so high that the plant would not be commercially competitive even with nil investment costs. According to a forecast by the owner/operator in 1973, operating costs are such that the net returns remaining from the sale of electricity are just enough to pay little less than the usual depreciation and interest on a capital of DM100 million. This was about the amount to be invested in the reactor core (fabrication and plutonium), which for the SNR-300 is paid for by the state, but in commercial operation would have to be financed as a fixed cost by the utilities. Although uranium prices have risen since then, the outlook on the fuel-cycle cost of the SNR-300 has worsened in recent years. Estimated fuel-fabrication costs are now twice as much (in constant money) than assumed in 1973, and estimates for repro-cessing costs have also increased considerably.

The high costs of the fuel cycle of the SNR-300 plant are partly due to the small scale on which all fuel-cycle facilities operate. Fabrication of fuel elements and reprocessing of spent fuel is to be done in small batches, so even small capacities cannot be utilized to the full. With construction of additional fast breeder plants the scale of the fast breeder fuel cycle increases and its economics are bound to improve. However, there are considerable technical and licensing uncertainties for all parts of the fast breeder fuel cycle, which render estimates of future fuel-cycle costs specula-tive.

Fuel fabrication for fast breeders, because of the toxicity of plutonium, has to be done in special plants. Although these plants can also be used to fabricate plutonium fuel for LWRs, the necessary resetting and cleaning of the plant between different types of fuel implies costly shutdown periods which, for small batches, are often longer than the respective production periods. Thus, operation in small batches incurs great penalties in terms of fixed plant costs. As of 1972, fabrication of the SNR-300 fuel (core and axial blanket) was estimated at ten times the fabrication cost of uranium fuel for LWRs. The fabrication experience up to now has necessitated a drastic upward revision of this estimate, while fabrication costs for LWRs remained stable over the last decade. As of 1979, estimates for SNR-300 are about thirty times higher than fabrication costs of uranium fuel for LWRs. If a larger number of fast breeders is put into operation, fuel fabrication will benefit from economies of scale. At present, West German utilities expect fabrication costs to come down to about 30 percent of SNR-300 cost or ten times those of LWRs.[19]

West German and Belgian manufacturers of plutonium fuel indicated that FBRs will not have fuel-cycle costs competitive to LWRs before the fuel for 3,000 to 5,000 MWe of fast breeder capacity is supplied by one fuel-fabrication line.[20] In the West Germany-Benelux area there are two fuel manufacturers, Alkem and Belgonucléaire. This means that between four and eight plants of the size of the SNR-2 will be necessary before both suppliers can produce at costs they regard consistent with competitive operation of fast breeder plants.

However, there are large uncertainties as to future economies of scale. A considerable development effort will be necessary in order to construct large-scale fuel-fabrication facilities. Their costs cannot be foreseen with certainty, as they will be critically dependent on licensing requirements. To improve the availability of future plutonium-fabrication plants, West German and Belgian suppliers are considering a modular-design concept.[21] Each module would have about the size of the present prototype facilities. This would mean that economies of scale would be limited.[22]

Blanket elements for fast breeders can be fabricated in the same facilities as the enriched-uranium fuel for LWRs. Uncertainties in terms of costs are, therefore, smaller than for fuel elements. For the SNR-300 the fabrication costs of blanket elements per kilogram of heavy metal are about three times those of LWR fuel. Although scaling up associated with standardization and concentration of fabrication will reduce costs in the future, utilities estimate that, because of differences in design and material, blanket elements for fast breeders will always have 50-percent higher costs per kilogram of heavy metal than LWR fuel.[23] As blanket elements will have a low burnup relative to LWR fuel, this implies that a notable trade-off between breeding and electricity costs will remain effective even after a full-scale fast breeder fuel cycle is established.

Transportation of spent fast breeder fuel is associated with specific problems due to its high decay heat and intensive radiation. There is an economic trade-off between the time that spent fuel is stored before shipping and the costs of transportation, as decay heat and radiation decrease with time, and less cooling and shielding is then required during shipping. It is reckoned that 100 days are the minimum storage time. Radiation and decay heat are then at the border of what is manageable with lead shielding and liquid-metal cooling. A storage time of 300 days or more will allow cooling by simpler means. Thus, long storage times make shipping cheaper, but they increase fixed fuel-cycle costs (the interest to be paid on the capital bound in the fuel cycle) and prolong the doubling time. For the initial large fast breeder plants there will be sufficient plutonium available from conventional nuclear plants, so the economic optimum will be with longer storage times.

The greatest uncertainties are connected with reprocessing and disposal of radioactive wastes. This service is not yet performed in a truly commercial framework even for LWR fuel, so there is little guidance for estimates. Originally it was planned to reprocess the SNR-300 fuel in the WAK proto-type facility at Karlsruhe. The estimated costs for reprocessing in this facility are not representative for future FBRs, as the WAK is a government facility and its investment costs including part of the costs of adapting the facility to fast breeder fuel were supposed to be covered by the government. This means that the reprocessing costs assumed in the forecast for SNR-300 operating costs are below what has to be expected in a commercial framework. Current negotiations with foreign fuel reprocessors indicate much higher costs.

For the first large fast breeders, reprocessing can be done either in commercial facilities designed for LWR fuel, or in special plants for fast breeder fuel. Because of the high plutonium content and the radioactivity of spent fast breeder fuel, it can be taken by reprocessing facilities for LWR fuel only if these are equipped with a special head-end, in which the fuel is mixed with blanket material; or if the whole facility operates with more diluted solutions than for LWR fuel. This means that reprocessing of fast breeder fuel incurs some cost penalties against LWR fuels.

There is an economic trade-off between the time that spent fuel is stored before reprocessing and the costs of reprocessing, similar to that in transportation. If spent fuel is stored before reprocessing, its radioactivity decreases and reprocessing can be done more cheaply. At the same time, however, doubling time becomes longer. The West German reprocessing plant for LWR fuel that is planned for a site in Hessen is designed for a cooling time of six years. Shorter cooling times, which are essential for the fast breeder, may imply considerable cost penalties. Another trade-off is between plutonium losses in reprocessing and the costs of reprocessing. It is assumed that 1 to 2 percent of the plutonium is not recovered from spent fuel and is disposed of with the radioactive wastes. Such losses not only worsen fuel-cycle economics but also prolong the doubling time, and make the disposal of highly active waste more difficult. The loss rate may be reduced by appropriate engineering, but at present it is not yet known what engineering efforts and what costs are to be traded against what reductions in the plutonium-loss rate. [24]

Uranium Resources

If LMFBRs are not competitive to LWRs at present uranium prices, they may become so in the future, as resources of low-cost uranium ores will be used up and more expensive uranium ores will have to be mined. The critical questions then are what price increase in uranium will be needed to tip the balance in favor of the breeder, and for how long uranium resources will allow an economic use of LWRs until the LMFBR becomes competitive.

Recent official West German justifications of the fast breeder program are based on the expectation that a global shortage of low-cost uranium may necessitate deployment of FBRs on a large scale before the year 2000.[25] This appears to be consistent with supply and demand projections for uranium by international organizations. A joint Working Party on Uranium Resources of the International Atomic Energy Agency (IAEA) and the OECD Nuclear Energy Agency (NEA) estimates uranium reserves (defined as reasonably assured resources up to a cost of $80 per kilogram of uranium) at 1.85 million tonnes.[26] According to the demand projections by the International Nuclear Fuel Cycle Evaluation (INFCE), this may satisfy demand up to 1998 or 2006, depending on assumptions on the growth of nuclear electricity generation and the extent of plutonium recycle.[27]

The economics of the FBR depend critically on the rate at which new uranium reserves are discovered and on the extent to which projections for nuclear generating capacities and the associated uranium demand materialize. Mineral reserves are generally defined in relation to costs of extraction, the state of technology, and the degree of certainty about the location and quality of deposits. Only those deposits can be economically mined, whose costs of extraction are consistent with the price at which the mined fuel can be sold. An increase in the price level, therefore, will allow more deposits to be mined economically. According to information assembled by the IAEA-NEA Working Party, 0.74 million tonnes uranium would be added to reserves if uranium prices rose to $130 per kilogram of uranium.

The probability that further exploration will increase reserves can be judged with some degree of confidence from present knowledge about resources, whose existence has been assured by prospecting and preliminary exploration but have not yet been sufficiently explored for consideration as reserves. The IAEA-NEA Working Party gives a figure of 1.48 million tonnes uranium in this category for a uranium price of up to $80 per kilogram, with an additional 0.97 million tonnes in the $80 to $130 price range.

More important for an assessment of future uranium is a consideration of the economic dynamics of prospection and exploration. Data on reserves are a function of past investments for the search of deposits. With continuing prospection and exploration, more deposits are likely to be identified as economically feasible and will then add to reserves. As prospection and exploration are mainly performed by business enterprises, they are guided by the firms' expectations for markets, sales, and returns on their investment. Usually the planning horizon of the firms for investments in exploration is ten to fifteen years in advance of anticipated demand. There are no economic incentives for a longer planning horizon for exploration. Firms do have longer planning horizons for prospection activities, as the costs of prospection are small compared to exploration.

In the uranium mining industry, anticipation of demand is not difficult within the time horizon which is relevant for the companies' investment decisions.[28] From start of detailed exploration to production of uranium concentrate and fabrication of reactor fuel takes about eight years. This is about the construction time for nuclear power plants. Uranium producers have therefore guidance for their investment activities from the plant-construction schedules of the utilities. As the investment in opening a mine is only worthwhile if the ore deposits allow operation for at least ten years, the desirable reserve situations is therefore defined by uranium producers as having assured uranium reserves about ten times larger than the yearly requirements eight years later. A comparison of the 1979 estimates of uranium reserves by the IAEA-NEA Working Party with the projections of yearly uranium demand in 1987 by INFCE shows a reserve situation much better than what is desirable according to the above definition. Using the most optimistic assumption on future nuclear capacities, yearly requirements in 1987 would be about 70,000 tonnes uranium, which would define a desirable reserve situation of 700,000 tonnes uranium. Reasonably assured resources up to $80 per kilogram of uranium were in 1979 at 1,850,000 tonnes uranium, that is, over twice as much. This is a factor large enough to cover economic and technical constraints that are neglected in this aggregate consideration.

The experience of the uranium industry suggests that through further prospection and exploration more uranium deposits will be found that can be economically exploited.[29] Most uranium reserves known today were prospected in the late 1940s and early 1950s, when the uranium industry experienced a boom as a result of high demands for military uses. Successful exploration in the early 1950s and the diminution of military requirements led to a period of overcapacity and low prices from 1959 to 1966 (see figure 7-1). A good number of mines were closed, and prospection activities ceased, until the first orders for commercial nuclear power plants started a new boom from 1966 to 1969. Another depression followed from 1970 to 1973, as reactor-construction schedules were delayed, and utilities became reluctant to engage in long-term commitments for fuel supply because of accumulating plant orders. Uranium prospecting experienced a strong push when oil prices increased in 1973–1974, and the uranium industry enforced a substantial price increase for uranium in the wake of the energy crisis.

Although the history of uranium prospecting and exploration has in general been one of boom and slump, additions to reserves were in the last decade larger than the amounts mined, so reserves grew considerably. From 1965 to 1973 reasonably assured resources (up to $39 per kilogram of uranium) increased by 700,000 tonnes, while 140,000 tonnes were mined over this period. This means an average addition of 105,000 tonnes uranium per

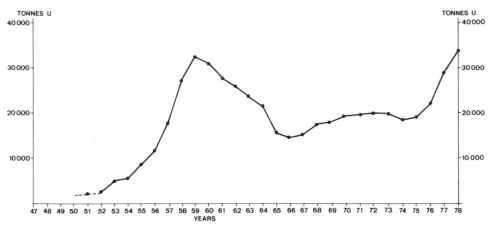

Sources: J. Cameron, "The Development of Uranium Resources: A Review of Some Problems Related to Uranium Supply for the Period 1975-2025," Paper EP/SEM.3/R 78, Economic Commission for Europe, Symposium on the Role of Electric Power in Meeting Future Energy Needs and on International Cooperation in This Field, Delphi and Athens, 19-23 May 1975, figure 1. Reprinted with permission. Data for 1974 to 1978 are from OECD Nuclear Energy Agency and International Atomic Energy Agency, *Uranium. Resources, Production and Demand* (Paris: Organisation for Economic Co-operation and Development, 1979), p. 22.

Figure 7-1. Uranium Production in the Western World, 1950 to 1978

year in a period during which the demand and price for uranium was low and firms had little incentive to invest in exploration. After 1973 additions to reserves have been on the same order of magnitude, though precise figures are difficult to obtain because of changing prices and definitions. [30]

In addition to past experience, there are geological indications that more uranium deposits will be identified in the future. As uranium has a very short history of prospection and exploration, there are vast areas not yet searched for uranium. According to one estimate, only 15 percent of the earth's surface has been prospected for uranium; another 25 percent is deemed geologically interesting, politically and environmentally available, and logistically accessible. [31] On the basis of indirect geological evidence, the IAEA-NEA Working Party expects that there are speculative uranium resources of 6 million to 15 million tonnes uranium with a cost up to $130 per kilogram. [32]

This will not limit uranium mining, if resources beyond $130 per kilogram are tapped. Low-grade uranium deposits, comprising phosphates, shales, and granites which can be mined and milled with present technology, though at a high price, contain around 50 million tonnes uranium according to an estimate by a West German government institute for geology. [33] The economic use of these resources seems less a matter of price than of environmental factors, as an increase in the price of uranium has a very small impact on the economics of present LWRs. The costs of natural uranium account for about 10 percent of total electricity costs in light-water plants.[34] This means that a doubling of the uranium price increases electricity costs by about 10 percent. Present commercial nuclear plants, therefore, seem able to stand quite high uranium prices without becoming noncompetitive with alternative sources of energy.

In the long run, advances in technology are likely to improve the efficiency of uranium extraction, and thus may allow mining deposits which cannot be exploited with present mining technology. There are about 4,000 million tonnes of uranium in oceanic waters at a concentration of about three milligrams per tonne. Uranium extraction from seawater is being done on a laboratory scale.[35] Its large-scale application, however, may or may not become economically feasible in the next few decades.

Uranium is one of the most abundant metals in the earth's crust. A West German study estimates that a total of 3.3×10^{12} tonnes exist down to a depth of 3,000 meters.[36] Because of low concentration, there are technical and environmental problems in tapping these hypothetical resources.[37] Uranium ores included in these hypothetical resources contain less than 50 grams of uranium per tonne. The question arises whether, at such low concentrations of uranium in rocks, the energy input in mining and milling will not be greater than the energy output as achieved in present commercial reactors. However, studies indicate that this is not a problem at the concentrations considered here.[38] Thus it is a matter of speculation what part of these hypothetical resources may be extracted some day.

Nevertheless, over the next 50 or 100 years, identification of new reserves does not seem to be a problem of resource depletion, but one of economic incentives. Uranium-producing companies do not feel constrained by a lack of favorable areas for uranium prospection, but by a lack of assurance about future demand for uranium. Although the price increases in recent years have improved prospects for returns on investments in uranium mining and the level of prospecting and exploration activities has responded, the uranium industry hesitates to make large investments because of fears that, through slippages in construction schedules of nuclear plant or cancellations, anticipated demand for uranium will not materialize.[39]

The experience of recent years substantiates this concern, as official projections of nuclear power growth have been steadily downgraded (see table 7–2). Even the events of 1973 and the resulting announcements of governments for increased reliance on nuclear power have not reversed the downward trend. At present, a change in this trend is not apparent, and current forecasts of future nuclear power capacities may still be too high. A recent market survey by the West German firm Nukem gives for the year

Table 7-2
Estimates of World Nuclear Capacity

Date of Estimate	World Nuclear Capacity (GWe)						
	1970	1975	1980	1985	1990	2000	2020
1969	25.6	101–125	235–330	—	—	—	—
1970	18	118	300	610	—	—	—
1973	14	94	264	567	1,070	—	—
1975	—	69	179–194	479–530	875–1,004	2,005–2,480	—
1977–1978	—	—	146	178–368	504–700	1,000–1,890	2,157–6,650
1979	—	—	144–159	244–272	374–460	834–1,207	1,800–3,900

Sources: Estimates up to 1973 are from R. Krymm and G. Woite, "Estimates of Future Demand for Uranium and Nuclear Fuel Cycle Services," *IAEA Bulletin* 18, no. 5/6 (1976):6–15, reprinted with permission. Other estimates are from OECD Nuclear Energy Agency and International Atomic Energy Agency, *Uranium. Resources, Production and Demand* (Paris: Organisation for Economic Co-operation and Development, 1975 and 1977); OECD Nuclear Energy Agency, *Nuclear Fuel Cycle Requirements and Supply Considerations, through the Long Term* (Paris: Organisation for Economic Co-operation and Development, 1978); International Nuclear Cycle Evaluation, *INFCE Summary Volume* (Vienna: International Atomic Energy Agency, 1980), p. 4.

Note: Estimates from 1969 to 1977 were made jointly by the OECD Nuclear Energy Agency and the International Atomic Energy Agency. The 1979 estimate is by the International Fuel Cycle Evaluation. All estimates are for world outside centrally planned economies.

2000 a world nuclear capacity of 754 GWe.[40] This is below the low case of the projection by INFCE.

If the low-growth projection by INFCE is accepted as realistic, the 5 million tonnes of uranium in the reasonably assured and estimated additional resource category may satisfy world demand roughly up to the year 2025; and estimated potential resources of 6 million to 15 million tonnes may carry nuclear power to the middle of the next century (see table 7-3).

The INFCE calculations also show that the uranium savings to be achieved by FBRs in the first few decades after their market introduction are highly sensitive to the growth rate of nuclear power. With once-through LWRs and a low growth rate of nuclear power, cumulative uranium demand up to 2025 will be in the same range as with large-scale use of FBRs and a high growth rate of nuclear power. In other words, a reduction of the compound growth rate of nuclear power from about 7 to 5 percent will make the same difference as having or not having the fast breeder.[41]

In view of the lengthy time delay before the fast breeder may show an economic benefit, an uncertainty exists whether other technologies will then be economically more attractive such as nuclear fusion, thermal breeder reactors, or novel technologies utilizing energy from solar radiation, winds, tides, waves, and thermal gradients of oceans. Over the long run, there may also be considerable technical improvements in the LWR fuel cycle, in addition to advances in mining and milling uranium. New enrichment technologies seem possible, that at the same time may improve the fuel utilization and decrease electricity costs.[42] Some of them, like laser enrichment, may pose proliferation problems.[43] But if we assume that a regime of non-proliferation policy is established in the long run that allows the large-scale use of FBRs, it is not unrealistic to expect that such a regime may also be found for advanced enrichment technologies.

If uranium prices rise, it may become economically attractive to operate LWRs (or heavy-water reactors) at higher conversion factors that significantly improve fuel utilization. A recent West German study proposed an advanced core design for a pressurized-water reactor (PWR), with a conversion ratio of about 0.95 and a five times better fuel utilization than achieved with present designs.[44] Although this is far below the fuel utilization of fast breeders, this technology may extend the commercial use of LWRs by some decades, and possibly even longer.

It follows from these reflections that there are great uncertainties about the fuel savings to be achieved through introduction of fast breeders. There are technical uncertainties with regard to the feasibility of the short out-of-core time and the low plutonium-loss factor required for a breeder with a good breeding potential. Economic uncertainties exist about trade-offs between breeding and electricity costs. There are uncertainties in the fore-

Table 7–3

Cumulative Natural Uranium Requirements for Illustrative Fuel-Cycle Strategies

Reactor Strategy	Nuclear Power Growth[a]	Uranium Requirements (million tonnes uranium)	
		2000	2025
LWR once through	High	2.0–2.2	9.5–12.1
	Low	1.5–1.6	5.4– 6.8
HWR once through	High	2.0	8.8–10.0
	Low	1.5	5.2– 5.8
LWR recycle	High	1.7–1.8	7.2– 9.0
	Low	1.3	4.0– 5.0
HWR recycle	High	1.9	7.9– 8.7
	Low	1.4	4.7– 4.9
FBR moderate use[b]	High	1.9	7.8– 8.4
	Low	1.3	4.6– 4.9
FBR large-scale use[c]	High	1.8	4.7– 6.9
	Low	1.3	3.5– 3.7

Source: International Nuclear Fuel Cycle Evaluation, *INFCE Summary Volume* (Vienna: International Atomic Energy Agency, 1980), p. 66. Reprinted with permission.

Note: World outside centrally planned economies.

[a] See table 7–2.

[b] Mixed strategy 3 in INFCE terminology.

[c] Single-type strategy 3 in INFCE terminology.

casts on the contribution of nuclear power to future electricity supply. And technical advances in the light-water fuel cycle may render fast breeder benefits imaginary for the next few generations.

Security of Supply

Energy is a vital input for the modern industrial economy. Interruptions of supply may cause high costs. World trade in fuels is highly oligopolistic. In some sectors it is dominated by supplier cartels. A country that is dependent on energy imports, like West Germany, engenders economic risks of supply interruptions and sudden price increases, or political risks of coercion by supplier nations. In the wake of the oil crisis of 1973 security of supply has received increased consideration, and governments have taken steps to foster development of indigenous energy supplies.

After the economics of the fast breeder have proved disappointing, proponents now turn to security of supply as the main argument for an early introduction of FBRs for commercial use. As a fast breeder plant is virtually independent of uranium supply, it is argued that it has to be considered as a domestic energy source and therefore has to be placed on the same footing as domestic coal, wind, and solar energy, or energy conservation. Domestic coal is subsidized in West Germany at a rate of DM4,000 million per year. The development of solar and wind energy and of technologies for improved energy use is funded at a rate of DM100 million per year. Private investments for more efficient heating are subsidized at a rate of DM1,100 million per year. And plans are discussed for a large subsidized program for coal liquefaction and gasification.[45] If government values security of supply so highly that it allocates these large subsidies, advocates of the FBR deem it logical that government subsidizes also the industrial use of FBRs in order to make the country independent of uranium imports. The 12,000 tonnes of uranium in West German soil are equivalent to one-year's demand of the West German nuclear program as projected for the year 2000.

FBRs are indeed a sort of insurance against uranium shortages and supply interruptions. But one must clearly distinguish which risks are effectively covered by this insurance and which are not. In an electricity economy like that of West Germany, it will take about five decades from the time the first commercial FBR comes on line until all nuclear plants are replaced by FBRs and independence of uranium imports is achieved. During this introductory period, the addition of new FBRs is limited by the plutonium produced in LWRs and from previously installed FBRs. The start of FBRs with uranium is feasible, but this would increase initial uranium demand. FBRs therefore provide security only for long-term supply. As long as they do not make up a significant fraction of nuclear generating capacity, they add little insurance against short-term shortages that might be created by a cartel of mining firms or by a coalition of foreign governments.

The industrial use of FBRs presupposes a fully developed fuel cycle for LWRs. This fuel cycle is in any case affected very little by short-term interruptions of up to two years or so. There will be large stocks of spent fuel which can yield uranium and plutonium, and there will be operating stocks at all facilities in the fuel cycle. Fuel utilization can be improved with small cost penalties through stretch-out operation or through increased burnup. If additional insurance is desired against short-term supply interruptions, other measures such as stockpiling, mining low-grade domestic ores, or extracting uranium from seawater may provide it at less cost than a series of subsidized large FBRs. In the longer term, advanced core designs that can be fitted into already operating LWR plants may provide as good an insurance as FBRs.[46] The advanced LWR cores would not have to be deployed before they were actually needed and would thus have a relatively low cost to be paid as insurance premium.

With regard to long-term uranium shortages that may result from exhaustion of low-cost uranium ore, a distinction is to be made between the insurance provided through the existence of a technological option and that provided through a subsidized large-scale construction program. It is good to have the FBR as a technological option to fall back on in the long term. But security of supply in the long term does not require a subsidized large-scale construction program of uneconomic FBR plants. As high-grade uranium ores become scarce, the price of uranium will rise, and uranium from lower-grade ores is likely to be available in sufficient quantity for a transitional period until FBRs will have replaced current nuclear technologies. For West Germany, the know-how to be acquired from the SNR-300 plant, together with continuing background R&D on critical technical problems in government research laboratories and ongoing industrial work on improved types of fast breeder fuels, seems to provide a sufficient level of insurance against these long-term risks. Instead of subsidizing a series of uneconomic FBRs with government money, this money would probably buy better insurance against long-term scarcity of energy supplies if it were spent to explore the feasibility of technical alternatives to the FBR. These may be nuclear technologies such as advanced converters with or without the thorium cycle or nonnuclear technologies.

For a uranium-importing country like West Germany these considerations on security of supply of nuclear fuels presuppose an international trade in uranium. Recent political actions and discussions surrounding the issue of military implications of nuclear electricity generation have created in uranium-importing countries a concern that uranium supplies might be subject to political restrictions that prevent an international market. However, for any workable regime of international nonproliferation policy the assurance of uranium supplies is a key element.[47] As long as a workable nonproliferation regime exists, there will be little economic incentive for increasing security of supply through subsidized large-scale use of FBRs. An exploratory R&D effort on FBRs together with the other measures mentioned earlier appears to provide sufficient insurance against the economic risks of a potential collapse of the present nonproliferation regime.

Export Potential

The West German economy is highly enmeshed in international trade. The need of sustaining a high performance in the export of manufactured goods is a standard argument for government support of R&D.[48] When it came to construction of the SNR-300, this argument was used (among others) to make a case for having an independent capability to manufacture FBRs as against the alternatives of importing FBRs, or taking out foreign licenses, or joining foreign programs as a junior partner. The official justification for the construction of the SNR-300 said that this plant must be built because the export of power plants plays an important role in world trade, and West German industry together with its Dutch and Belgian partners must demonstrate that it can master the advanced technology of FBRs.[49]

Despite the great attention that exports of nuclear plants receive at the political level, a look at available statistics suggests that it is not as important in relation to total exports. Although West German nuclear industry has done relatively well on the world market, exports in nuclear technology in 1978 accounted for about 0.2 percent of total exports. As table 7-4 shows, the fraction of exports on total nuclear capacity manufactured is higher in West Germany than in other countries. The export ratio of nuclear industry (38.8 percent) is above the average of West German capital goods industry (34.1 percent).[50] Nevertheless, exports of nuclear reactors and of parts thereof were in 1978 at DM17.0 million (see table 7-5) or 0.006 percent of total West German exports. Imports in reactor parts were at DM76.6 million so there was a net balance of trade deficit. Of course the export of nuclear plants involves a great many items in addition to the reactor itself: turbine generators, switch gear, instrumentation, and fuel elements. These appear in other titles of the trade statistics. On the other hand, to some extent these items can be exported independent of the sale of whole nuclear plants. Thus the trade statistics in table 7-5 need cautious interpretation. Exports of nuclear primary fuels may be disregarded, as these are reexports. If the remaining subtotal in table 7-5 is taken as an indicator for West German exports in nuclear technology, these account for 0.2 percent of total exports.[51]

A look at the world market in nuclear power plants suggests that the role of exports of nuclear plants will not drastically increase in the future. Available figures on nuclear generating capacity sold up to the present and projections on future nuclear capacity suggest that the major part of the market is in industrially advanced countries which have their own nuclear industry (see table 7-6). There is only limited scope for international trade in nuclear plants among these countries. Utilities have a preference for domestic suppliers, especially where they are nationalized.[52] The usually highly concentrated power-plant markets can be easily protected by governments which have an interest in a potent domestic nuclear industry. The socialist countries are an uncertain market for Western suppliers. Apart from one plant in Yugoslavia, all plants in socialist countries have been supplied by the Soviet Union. China may become an importer of Western nuclear plants, but the future role of nuclear power in China is difficult to assess.

Thus the potential for exports by Western manufacturers is basically limited to two groups of countries: the advanced industrial countries which

Table 7–4
Exports of Nuclear Plants in the Western World

Plants in Operation, under Construction, on Order, and Decommissioned as of 31 December 1979

Country	Exports		Domestic		Total		Exports as Percentage of Production
	Number of Plants	GWe	Number of Plants	GWe	Number of Plants	GWe	
United States	62	43.7	202	196.3	264	240.0	18.2%
West Germany	16	16.2	33	25.6	49	41.8	38.8%
France	8	6.9	67	64.8	75	71.7	9.6%
Canada	5	2.2	26	15.6	31	17.8	12.4%
Sweden	2	1.3	10	7.9	12	9.2	14.1%
Britain	3	0.5	41	13.0	44	13.5	3.7%
Japan	—	—	19	12.8	19	12.8	—
Italy	—	—	5	3.9	5	3.9	—
Others	—	—	8	1.1	8	1.1	—
Total	96	70.8	411	341.0	507	411.8	17.2%

Source: "atw-Schnellstatistik: Kernkraftwerke 1979—Weltübersicht," Atomwirtschaft 25 (1980):157–162.
Note: Small differences to table 7–5, because table 7–5 does not include decommissioned plants.

Table 7-5
West German Exports and Imports in Nuclear Technology
(DM million)

	Exports			Imports			Balance		
	1976	*1977*	*1978*	*1976*	*1977*	*1978*	*1976*	*1977*	*1978*
Nuclear reactors and parts	33.5	54.5	17.0	33.8	26.1	76.6	−0.3	+28.4	−59.6
Unirradiated fuel elements	87.6	146.3	497.5	5.9	30.5	47.6	+93.5	+115.8	+449.9
Other technical products[a]	74.0	96.7	81.2	82.9	86.7	76.7	−8.9	+10.0	+4.5
Radioactive material and isotopes	68.8	36.5	36.7	49.8	57.5	64.5	+19.1	−21.0	−27.8
Subtotal	263.9	334.0	632.4	172.4	200.8	265.4	+91.5	+133.2	+367.0
Nuclear primary fuels	121.3	252.1	327.0	546.7	891.4	1,314.8	−425.4	−639.3	−987.8
Total	385.2	586.1	959.4	719.1	1,092.2	1,580.2	−333.8	−506.1	−620.8
Total as percentage of all West German exports	0.15	0.21	0.34						

Source: Statistisches Bundesamt, *Aussenhandel nach Waren und Ländern (Spezialhandel)*, Fachserie 7, Reihe 2 (Stuttgart and Mainz: W. Kohlhammer, 1977–1979, annual); "Der Aussenhandel der Bundesrepublik Deutschland mit kerntechnischen Erzeugnissen 1976 bis 1978," *Atomwirtschaft* 24 (1979):535.

[a] Facilities for radiation, radiation protection, particle accelerators, and machines for the handling of radioactive materials.

Table 7-6
The World Market in Nuclear Power Plants

Countries	Nuclear Capacity in Operation, under Construction, or on Order as of 1979		Nuclear Capacity as Projected by INFCE for the Year 2000	
	GWe	Percentage of Total	GWe	Percentage of Total
Advanced Western countries with own nuclear industry [a]	400	79.4	627–937	57.8–56.7
Advanced Western countries importing nuclear plant	24	4.8	58–81	5.4–4.9
Centrally planned economies	48	9.5	248–447	22.9–27.1
Developing countries	32	6.3	151–187	13.9–11.3
Total	504	100.0	1,084–1,652	100.0

Sources: "Kernkraftwerke 1979—Weltübersicht," *Atomwirtschaft* 25 (1980):157–162; International Nuclear Fuel Cycle Evaluation, *Fuel and Heavy Water Availability*, Report of INFCE Working Group 1 (Vienna: International Atomic Energy Agency, 1980), pp. 44–47.

[a] Includes United States, France, West Germany, Japan, Canada, Britain, Sweden, Belgium, and Italy.

have not built up a domestic nuclear industry, and the developing countries. These two groups together account at present for only 11 percent of the nuclear world market. Their imports of nuclear plants may increase in the future, as demand for electricity grows. But this will probably not increase their share on the world market.

Developing countries are inhibited by a number of factors in the large-scale use of nuclear power.[53] A market for nuclear plants is likely to exist in the near future only in the larger of these countries, as smaller countries may find it difficult to accommodate plants of a unit size of 1,000 MWe and beyond. To keep risks of unavailability within limits, the largest plant in a national grid must not contribute more than 5 to 10 percent to total generating capacity. This makes large nuclear units unattractive at present for most developing countries. Small nuclear plants with a capacity below 600 to 800 MWe have higher capital costs per kilowatt of electrical output and, therefore, higher electricity costs. The high capital costs of nuclear plants will strain the financial capability of developing countries. Considerations of balance of payments and of opportunity costs will make the less capital-intensive use of fossil or solar energy attractive, wherever fossil or solar sources are available. In developing countries there is a limited base of experience with licensing, building, and operation of nuclear plants.[54] In many of these countries, this will render a rapid expansion of nuclear power unlikely.

Some of these problems may be overcome within the next two or three decades. However, for a longer time horizon there are uncertainties in the projected growth of electricity demand. Most forecasts of electricity demand in developing countries are guided by the patterns of economic growth and energy use that evolved in industrially advanced countries in periods of low-cost energy. It is questionable that with oil prices several times higher developing countries will follow the same growth path.

The export potential of FBR plants will be more restricted than that of LWRs. FBRs are more capital-intensive. Developing countries with limited investment capital and scarce foreign currencies will therefore prefer the less capital-intensive technologies, even if the fast breeder should become commercially attractive. The unit size of commercial fast breeder plants will be around 1,500 and possibly 2,000 MWe. They will therefore require national electricity-generating capacities two times larger than present LWRs. Several countries may be capable of accommodating 1,000-MWe units, but 2,000-MWe units are beyond their reach. Fast breeders are a more exacting technology to license, construct, and operate. There would be therefore little wisdom in ordering such plants before a country has acquired a large base of experience in thermal reactors. For countries without a fully established FBR fuel cycle, the breeder will not provide independence in fuel supply. Without timely reprocessing of spent fuel and refabrication of plutonium into fuel elements, the breeder runs out of fuel and its economics go awry. Countries without facilities for reprocessing and refabrication will therefore be crucially dependent on foreign countries for these services.

Exports of FBRs will be limited by the plutonium produced in conventional nuclear plants. If the fast breeder should become economically attractive, exporting countries are likely to reserve their plutonium for domestic use. This plutonium will be in the hands of the utilities (which will need it for fueling their own breeders), not reactor vendors. Countries importing fast breeders will therefore have to rely on the plutonium produced in their own nuclear plants.

Problems of proliferation of nuclear weapons are more difficult for fast breeders than for present commercial reactors.[55] All types of present commercial reactors produce plutonium in amounts similar to or even larger than FBRs. The plutonium produced in present nuclear plants is not accessible before the spent fuel is reprocessed, because spent fuel is highly radioactive for a longer time. For present commercial reactors the recycle of plutonium is a possible technical option, but they can be economically operated without reprocessing and the use of plutonium fuel. Fast breeders, however, cannot be economically operated without recycle. The plutonium can be extracted from the fresh fuel by relatively simple chemical methods, and the quantitites are such that a single core load of fuel elements can supply explosives for dozens of bombs.

Furthermore, fast breeders can be operated in such a manner that they produce high-grade plutonium. The quality of the plutonium produced in a reactor is dependent on the burnup. With higher burnup the fraction of nonfissionable plutonium isotopes increases, which makes the plutonium less efficient though not unusable as an explosive. Weapon-grade plutonium can be produced in any type of reactor just by unloading the fuel elements after a low burnup. But this is unusual in normal operation, and can be detected by controls. In fast breeders, however, high-grade plutonium is produced in the breeding blankets and can be taken out in normal reload intervals.

Apart from all these considerations, fast breeders will have to prove their economic competitiveness before exports are possible. As we have seen before, prospects for the economic performance of fast breeders are dim. Mastering a sophisticated but uncompetitive technology will do little to improve the performance of West Germany's power-plant industry on the world market.

Conclusion

Provided that the safety and proliferation risks of FBRs can be reduced to acceptable levels, a decision strategy that makes the construction of future commercial-size demonstration plants contingent on the willingness of utilities and manufacturing firms to finance them from their own funds, seems to make proper accounting of the social benefits claimed for this technology. If the economic prospects of FBRs are such that the second or third plant after the SNR-2 would be economic, the enterprises involved, backed by government-financed exploratory R&D and risk sharing, would be capable of commercializing this technology with their own funds. If economic prospects are less bright, and this is what our analysis suggests, there are serious doubts that government subsidies for a series of uncommercial FBRs would be a better investment than exploratory R&D on technical alternatives.

Notes

1. For West Germany see table 3-1.

2. Interviews with participants, February 1975, March 1976, September 1976, October 1976, June 1977. The reluctance to criticize is nicely illustrated in an internal memo by a ministry official that is among the papers of the German Atom Commission. It records a telephone call by an industrial member in the Working Group, "Nuclear Reactors," complaining that out of a longer discussion a remark of his was recorded in the minutes that could be read as an indirect criticism of the project discussed. The member points out that this remark will be regarded as an unfriendly act by the firm which proposed the project, and this will get him into trouble with his own management (Federal Archives Volume B 138 3368, Ref. zu II D-1817-3/1-34/60).

3. In February 1974 an interviewee gave a figure of DM 320 million to DM330 million. This was inflated to June 1979 prices using the price index for industrial capital goods.

4. In the business year 1978-1979 Siemens spent DM2,700 million on R&D, of which about 90 percent was self-financed (Siemens AG, *Annual Report 1978-1979*).

5. Siemens's self-financed R&D on nuclear reactors up to 1967 is estimated in nuclear circles to be of the order of DM100 million (see chapter 2). Cumulative sales from 1955-1956 to 1966-1967 were DM62,497 million, sales in 1978-1979 were DM28,022 million.

6. LWRs cost in 1979 about DM2,000 per kilowatt, excluding inflation and interest during construction, but including owner's cost and the first core (A.-W. Eitz, "The Economic Interest of Fast Breeder Reactors," paper presented at the Council of Europe Parliamentary Hearing, Brussels, 18-19 December 1979). A 1,300-MWe fast breeder plant costing 150 percent of an LWR would thus have a capital cost differential of around DM1,300

million. RWE's sales in 1978–1979 were DM15,669 million (see the company's annual report).

7. The Versuchsatomkraftwerk Kahl/Main (VAK) was built in the years 1958 to 1961, at a cost of DM49 million, of which RWE financed 80 percent (see chapter 2). RWE's total sales in the period July 1958 to June 1961 were DM5,436 million (see the company's annual reports).

8. For other critical discussions of the economics of FBRs see I.C. Bupp and J.C. Derian, "The Breeder Reactor in the U.S.: A New Economic Analysis," *Technology Review* (July/August 1974):26–36; T.B. Cochran, *The Liquid Metal Fast Breeder Reactor: An Environmental and Economic Critique* (Baltimore and London: Johns Hopkins University Press, 1974); B.G. Chow, *The Liquid Metal Fast Breeder Reactor: An Economic Analysis* (Washington, D.C.: American Enterprise Institute, 1975); M. Sharefkin, *The Fast Breeder Reactor Decision: An Analysis of Limits and the Limits of Analysis*, prepared for the Joint Economic Committee, Congress of the United States (Washington, D.C.: U.S. Government Printing Office, 1976); Ford-Mitre Nuclear Energy Policy Group, *Nuclear Power: Issues and Choices* (Cambridge, Mass.: Ballinger, 1977); H.A. Feiveson, F. von Hippel, and R.H. Williams, "Fission Power: An Evolutionary Strategy," *Science* 203 (26 January 1979): 330–337; B.G. Chow, "Economic Comparison of Breeders and Light Water Reactors," report to the U.S. Arms Control and Disarmament Agency (Los Angeles, Calif.: Pan Heuristics, 1979).

9. As of June 1979, the SNR-300 is estimated by SBK to cost DM2,117 million in 1972 money, excluding owner's cost and the first core. If inflation up to mid-1979 is included the estimate is DM3,260 million. Per kilowatt of net electric output this is DM10,867 or DM11,644, depending on whether the design net output of SNR-300 (300 MWe), or the guaranteed net output (280 MWe) is assumed. Costs of West German LWR plants as of mid-1979 are estimated at DM1,800 to DM1,850, excluding owner's cost and the first core (interview with a participant).

10. Interview with a participant, October 1977.

11. M. Rozenholc, A. Brandstetter, and J. Moore, "Fast Breeder Reactor in Europe," *Proceedings ENC '79* (Bonn: Deutsches Atomforum, 1979), pp. 91–100. French cost data for nuclear plants tend to be considerably lower than American or West German data because of differences in cost definitions as well as in industrial structure; see M. Gauzit and R. Fuhrmann, "Le Côut des Centrales Nucléaires: Essai de Comparaison entre la France et les Etats-Unis," *Revue Générale Nucléaire,* no. 1 (1977):39–46. The same plant may cost in the United States 1.7 times more than in France. This must be taken into account when comparing the cost of Super-Phénix to an American or West German LWR.

12. I.C. Bupp et al., "The Economics of Nuclear Power," *Technology*

Review (February 1975):14–25; R. Krymm, "A New Look at Nuclear Power Costs," *IAEA Bulletin* 18, no. 2 (1976):2–11; W.E. Mooz, "Cost Analysis of Light Water Reactor Power Plants," Report R-2304-DOE (Santa Monica, Calif.: Rand Corporation, 1978).

13. M. Diedrichs, "Standardisierung von Kernkraftwerken mit Leicht-wasserreaktoren," *VGB Kraftwerkstechnik* 54 (1974):578–582.

14. A.J. Surrey, "The Future Growth of Nuclear Power," *Energy Policy* (1973):107–129, 208–224; M. Lönnroth and W. Walker, "The Viability of the Civil Nuclear Industry," working paper for the International Consultative Group on Nuclear Energy (New York: Rockefeller Foundation, London: The Royal Institute of International Affairs, 1979).

15. For example, M. Levenson, P.M. Murphy, and C.P.L. Zaleski, "Relative Capital Cost of the LMFBR," *Proceedings of the American Power Conference* 38 (April 1976):265–269; Battelle Columbus Laboratory, "Study of Advanced Fission Power Reactor Development for the United States," Report BCL-NSF C 946-2 (Columbus, Ohio: Battelle Columbus Laboratory, 1976).

16. W.E. Mooz and S. Sigel, "A Comparison of the Capital Costs of Light Water Reactor and Liquid Metal Fast Breeder Reactor Power Plants," Report R-2441-ACDA (Santa Monica, Calif.: Rand Corporation, 1979).

17. Assumed is a load factor of 0.7 and an annual fixed charge of 14 percent.

18. The following draws mainly on H. Schmale, "Probleme des Brennstoffkreislaufs Schneller Brüter," *Atomwirtschaft* 18 (1973):415–418, and on interviews with participants, December 1974, December 1975, March 1980.

19. Eitz, op. cit., note 6.

20. W. Stoll and E. van den Bemden, "Fuel Fabrication Problems for SNR 300," in *Status Bericht Schneller Brüter 1973,* edited by Schnell-Brüter-Kernkraftwerksgesellschaft und Internationale Natrium-Brutreaktor-Bau Gesellschaft, n.d., pp. 194–115.

21. W. Stoll, "Experience in Operating a Pu-Fabrication Facility," in Proceedings of the Joint Topical Meeting on Commercial Nuclear Fuel Technology Today, Toronto, Canada, 28–30 April 1975, sponsored by the American Nuclear Society and the Canadian Nuclear Association, pp. 4–24 to 4–43; W. Stoll, "Experience in Fabricating Plutonium Recycle Fuel," in *The Plutonium Fuel Cycle,* proceedings (Hinsdale, Ill.: American Nuclear Society, 1977), pp. III.2–1 to III.2–8.

22. For an optimistic view see J. van Dievoet et al., "The RNR Fast Breeder Reactor Fuel Cycle," paper read at European Nuclear Conference, Hamburg, 6–11 May 1979.

23. Schmale, op. cit., note 18, p. 417.

24. For some illustrations on the effect on doubling time of out-of-reactor time and plutonium losses, see R.L.R. Nicholson and A.A. Farmer, "The Introduction of Fast Breeder Reactors for Energy Supply," in *Uranium and Nuclear Energy,* proceedings of the Fourth International Symposium held by the Uranium Institute, London, 10–12 September 1979 (London: Mining Journal Books, 1980), pp. 141–168.

25. Bundesminister für Forschung and Technologie, "Bericht über die Entwicklung des Natriumgekühlten Schnellbrutreaktors an den Ausschuss für Forschung und Technologie und an den Haushaltsausschuss des Deutschen Bundestages," September 1977; see also M. Kempken et al., "Future Plans for the Design and Construction of Fast Reactor Power Stations in the Federal Republic of Germany," in *Design, Construction and Operating Experience of Demonstration LMFBRs: Proceedings of a Symposium, Bologna, 10–14 April 1978* (Vienna: International Atomic Energy Agency, 1978), pp. 775–783.

26. OECD Nuclear Energy Agency and International Atomic Energy Agency, *Uranium. Resources, Production and Demand* (Paris: Organisation for Economic Co-operation and Development, 1979).

27. International Nuclear Fuel Cycle Evaluation, *Fuel and Heavy Water Availability,* report of INFCE Working Group 1 (Vienna: International Atomic Energy Agency, 1980), p. 88.

28. H.-R. Hampel and A. von Kienlin, "The Supply of Uranium," reprint from *Metallgesellschaft AG—Review of Activities,* no. 16 (1973): 40–50.

29. J. Cameron, "The Development of Uranium Resources: A Review of Some Problems Related to Uranium Supply for the Period 1975–2025," Paper EP/SEM.3/R 78, Economic Commission for Europe, Symposium on the Role of Electric Power in Meeting Future Energy Needs and on International Cooperation in this Field, Delphi and Athens, 19–23 May 1975; J.S. Foster et al., *Nuclear Resources: The Full Report to the Conservation Commission of the World Energy Conference* (New York and Guildford: IPC Science and Technology Press, 1978), pp. 165–176.

30. International Nuclear Fuel Cycle Evaluation, op. cit., note 27, pp. 166–168.

31. M. Hansen, "Trends in Uranium Supply," *IAEA Bulletin* 18, no. 5/6 (1976):16–27.

32. OECD Nuclear Energy Agency and International Atomic Energy Agency, *World Uranium Potential* (Paris: Organisation for Economic Co-operation and Development, 1978).

33. Bundesanstalt für Geowissenschaften und Rohstoffe, "Das Angebot von Energie-Rohstoffen," report, March 1976.

34. H. Michaelis, "Ist Kernenergie wirtschaftlich?," KAT-3-77 (Bonn: Deutsches Atomforum, 1977); T. Roser, "The Economics of Nuclear

Power Generation," in *Foratom VII. National Reports. Hamburg, 6–9 May 1979* (Bonn: Deutsches Atomforum, 1979), pp. 45–46.

35. Bundesanstalt für Geowissenschaften und Rohstoffe, op. cit., note 33. S. Hayashi, "A Review of the Uranium Extraction from Seawater Plan in Japan," and F. Pantanetti, "The Operation of Recovering Uranium from Sea Water," both in *Nuclear Energy Maturity: Proceedings of the European Nuclar Conference, Paris, 21–25 April 1975,* vol. 11, edited by P. Zaleski (Oxford and New York: Pergamon Press, 1976), pp. 133–135; P.H. Koske, "Extraction of Urianum from Sea Water," in *Uranium and Nuclear Energy,* note 24, pp. 111–117.

36. Bundesanstalt für Geowissenschaften und Rohstoffe, op. cit., note 33.

37. H. Venzlaff, "Arme Uranerze als mögliche Rohstoffquelle," *Atomwirtschaft* 21 (1976):398–401.

38. E. Svenke, "Reasonable Limits for Low Grade Ores from Geological, Economic and Energy Balance Points of View," in *Nuclear Energy Maturity,* note 35, pp. 135–136.

39. For example, R.L. Dickemann, "Energy Self-Sufficiency: Fuel Cycle Bottlenecks," *Annals of Nuclear Energy* 2 (1975):745–749.

40. *NUKEM Market Report* (September 1979).

41. Similar projections on national uranium demand show the same result. Calculations for West Germany by proponents of the FBR show that a reduction by about two percentage points of projected annual growth rates of electricity demand may have the same effect on cumulative uranium demand up to the year 2040 as the substitution of FBRs for present types of nuclear plants. This is based on oxide-fueled fast breeder reactors, commercial introduction of FBRs by 1987, and electricity growth rates of 4.9 percent (low scenario) and 6.5 percent (high scenario) for 1980 to 1990, 5.1 and 5.3 percent, respectively, for 1990 to 2000, 2.3 and 4.6 percent for 2000 to 2010, 1.8 and 3.1 percent for 2010 to 2020, 1.4 and 2.0 percent for 2020 to 2030, 0.8 and 1.0 percent for 2030 to 2040; see R. Schröder and J. Wagner, comps., "Überlegungen zur Einführung Schneller Brutreaktoren im DeBeNeLux-Bereich," Report KFK-Ext. 25/75-1 (Karlsruhe: Gesellschaft für Kernforschung, June 1975).

42. P.R. Vanstrum and S.A. Levin, "New Processes for Uranium Isotope Separation," in *Nuclear Power and Its Fuel Cycle,* vol. 3, part 2 (Vienna: International Atomic Energy Agency, 1977), pp. 215–226.

43. A.S. Krass, "Laser Enrichment of Uranium: The Proliferation Connection," *Science* 196 (13 May 1977):721–731.

44. H.-H. Hennies and H. Märkl, "Überlegungen zur Modifizierbarkeit eines LWR in Hinblick auf bessere Uranausnutzung," in *Tagungsbericht der Jahrestagung Kerntechnik '80, Berlin, 25.–27. März 1980* (Bonn: Deutsches Atomforum, 1980), pp. 953–956. This concept uses the

plutonium cycle. Other advanced concepts for the LWR and the heavy-water reactor involve the use of thorium; see E. Critoph, "The Thorium Fuel Cycle in Water-Moderated Reactor Systems," in *Nuclear Power and Its Fuel Cycle,* vol. 2, part 1 (Vienna: International Atomic Energy Agency, 1977), pp. 55–74; J. Veeder, "Thorium Fuel Cycles in CANDU," *Transactions of the American Nuclear Society* 29 (1978):267–276; and W.B. Lewis, "New Prospects for Low Cost Thorium Cycles," *Annals of Nuclear Energy* 5 (1978):297–304.

45. Federal allocations for direct subsidies to coal were in 1979 DM1,850 million (*Siebter Subventionsbericht,* Bundestagsdrucksache 8/3097, August 1979); a further DM2,180 million were financed through a levy on electricity consumption (letter by Bundesminister für Wirtschaft, June 1980). For subsidies to new nonnuclear energy technologies see Bundesminister für Forschung und Technologie, *Programm Energieforschung und Energietechnologien 1977–1980* (Bonn: Bundesminister für Forschung und Technologie, 1977). For the program for support of energy-saving investments see *Zweite Fortschreibung des Energieprogramms der Bundesregierung,* Bundestagsdrucksache 8/1357, December 1977.

46. See Hennies and Märkl, op. cit., note 44.

47. T.L. Neff and H.D. Jacoby, "Supply Assurance in the Nuclear Fuel Cycle," *Annual Review of Energy* 4 (1979):259–311.

48. O. Keck, "West German Science Policy Since the Early 1960's: Trends and Objectives," *Research Policy* 5 (1976):116–157.

49. U. Däunert and W.J. Schmidt-Küster, "Das Projekt SNR—staatliche Förderung und internationale Zusammenarbeit," *Atomwirtschaft* 17 (1972):363–366.

50. Statistisches Bundesamt, ed., *Statistisches Jahrbuch 1979 für die Bundesrepublik Deutschland* (Stuttgart and Mainz: W. Kohlhammer, 1979), p. 174.

51. A general estimate arrives at the same order of magnitude. It may be assumed that the West German exports listed in table 7–3 stretch over a fifteen-year period; direct plant price is DM1,500 per kilowatt; 40 percent of supplies are contracted to firms in the importing country or in third countries. Then the annual value of these exports is DM972 million. This is 0.3 percent of total West German exports as of 1978.

52. A.J. Surrey and J.H. Chesshire, *World Market for Electric Power Equipment* (Brighton: University of Sussex, Science Policy Research Unit, 1972).

53. Some of the following considerations concerning less developed countries draw on a report by Richard J. Barber Associates to the U.S. Energy Research and Development Administration, "LDC Nuclear Power Prospects, 1975–1990: Commercial, Economic and Security Implications," February 1975, mimeograph.

54. M. Rosen, "The Critical Issue of Nuclear Plant Safety in Developing Countries," *IAEA Bulletin* 19, no. 2 (April 1977):12–21; M. Rosen, "Upgrading the Safety Assessment of Exported Nuclear Power Plants," in *Problems Associated with the Export of Nuclear Power Plants,* proceedings of a Symposium held by the IAEA in Vienna, 6–10 March 1978 (Vienna: International Atomic Energy Agency, 1978), pp. 27–42.

55. V. Gilinsky, "Fast Breeder Reactors and the Spread of Plutonium," Report RM-5148-PR (Santa Monica, Calif.: Rand Corporation, 1967); B. Barré, "The Proliferation Aspects of Breeder Deployment," in *Nuclear Energy and Nuclear Weapon Profileration,* edited by Stockholm International Peace Research Institute (London: Taylor & Francis, 1979), pp. 127–140.

8

Observations and Generalizations

The experience in the West German fast breeder program casts serious doubts on the economic justification for direct government subsidies to civilian technology. It suggests that a regime of government support that confines the government's role to background research and exploratory development, in some cases combined with loans for and risk sharing in demonstration plants, may be a better use of public money. This chapter reviews the findings with a view to evaluating the conflicting sets of arguments or theories presented in chapter 1. In doing so, it looks for possible generalizations that may be derived from the findings. Of course, a single case study does not yield a sufficient basis for firm generalizations. Nevertheless, some tentative generalizations or hypotheses may be put forward by asking whether the findings reflect idiosyncrasies of the actors involved or whether they can be traced to more general factors that are likely to be effective also in other government-financed R&D programs in West Germany and in other countries.

Influence

The relevant part of policymaking in the West German fast breeder program took place between the Federal Ministry in charge of the program, the organizations performing it, and the ministry's advisory bodies. In none of the cases analyzed in this study was a decision taken by the ministry significantly altered by other political institutions, such as the Finance Ministry, the cabinet, or parliament.[1] Until recently, none of these institutions had ever conducted a review of the fast breeder program, which would permit it to challenge the program's direction.

Parliamentary control was practically absent. The Bundestag committee in charge of government R&D policy drew its information mainly from the ministry, occasionally also from leading individuals from the organizations performing the program or from foreign institutions involved in fast breeder development. Even had it been willing, the Bundestag would simply not have commanded the information necessary to alter any of the decisions. In the period covered by this study the fast breeder never became an issue in party politics. Individual members of the Bundestag questioning the

223

wisdom of the program were rare exceptions and had no impact on policy-making.[2]

Party politics became involved only in the second half of the 1970s when a growing public opposition to nuclear power drew the attention of media and policymakers to the safety issues surrounding potential accidents in FBRs and to potential political implications of large-scale plutonium use.[3] The new American nonproliferation policy announced by President Carter in April 1977 calling for a delay of commercial reprocessing and the commercialization of FBRs triggered some political controversies in West Germany, which also concerned the SNR-300 and preparations for SNR-2. In September 1978, the Economic Minister of Nordrhein-West-falen, H.-L. Riemer (FDP), suggested changing the core design of SNR-300 to a plutonium-burner using thorium as fertile material.[4] The ensuing controversies finally led to a decision by the Bundestag to set up an inquiry, which was to discuss future nuclear energy policy and to recommend on whether the SNR-300 should by put in operation when completed. The inquiry issued an intermediate report in June 1980, but deferred a recommendation on the SNR-300 until its final report.[5] So far, party politics raised a good deal of discussion, but has not had an impact on the direction or the pace of the fast breeder program.

Over the period analyzed in this study, the press generally followed the views of the Karlsruhe laboratory or the ministry. The sole exception was Kurt Rudzinski of the *Frankfurter Allgemeine Zeitung*, who attacked the LMFBR since the mid-1960, favoring the steam-cooled FBR, and later the high-temperature gas-cooled reactor (HTGCR).[6] The rest of the press, however, regarded this criticism as something of a curiosity. It was taken as a subject to report on rather than an invitation to a public debate.[7]

Although the FBR represents the largest government-financed development program in the Federal Republic, the academic community has not yet regarded it as a challenge for independent analysis and evaluation.[8] The government has never commissioned institutions not directly involved in the program for an evaluation of it, and evaluations by advisory bodies, such as the ad hoc committee Support of SNR-300, were not published. Press releases by the ministry and the Karlsruhe laboratory were thus the main source of information for the general public.

Public opposition on environmental grounds was voiced when construction began on the SNR-300. In the first half of the 1970s it was confined in West Germany to local groups.[9] It was more widespread in the Netherlands, where for some time the Dutch part of the construction costs of this plant was paid through a 3-percent levy on electricity bills.[10] In the second half of the decade, public opposition grew considerably and led to a number of legal interventions in the licensing procedure. They culminated in the decision of an administrative district court in August 1977 to ask the judgment of the constitutional court on whether the licensing of FBRs is consistent with constitutional law. An amendment of the Atomic Law was prepared by the government, so as to avoid a construction delay of the SNR-300. But it was not needed, as the constitutional court decided in August 1978 that the construction of the SNR-300 is possible within existing law.[11]

This study has clearly shown that the dominant influence on the decisions in the West German FBR program was not from the industrial sector but from the government organizations, namely, the ministry in charge of the program and the Karlsruhe research laboratory. The idea for the fast breeder program was born at Karlsruhe. For the newly established research laboratory, the program was important in defining its function and in securing continuing employment for the team which had designed and constructed the FR2 reactor. The Federal Ministry for Atomic Energy received the initiative favorably, as it fitted well the ministry's strategy of expansion of the Karlsruhe laboratory, which indirectly was a strategy of consolidating the ministry's own responsibilities.

As a result of the vicissitudes of European intergovernment collaboration, the fledgling FBR program received a major push. Euratom was interested in supporting FBRs because it was largely precluded from the first generation of nuclear power plants: light-water reactors (LWRs) or gas-graphite reactors. Financing fast breeder actitivies carried the hope for a unified European industry and for a larger role of Euratom in what was then regarded as an inevitable second generation of nuclear plants. Instead of a unified European program, however, French and West German governments preferred national programs in affiliation with Euratom. This was in the interest of the West German government, since it was eager to see a fair return from its contributions to Euratom's budget. This policy required that the Karlsruhe program be accelerated and expanded. As a result, national and international funds were committed on a scale exceeding West German government subsidies to LWRs in the 1960s. For Euratom, finally, an association with national FBR programs was better than no role in fast breeders at all.

The initiative for an acceleration of the program in 1965–1966, including the start of the design of a steam-cooled and a sodium-cooled 300-MWe prototype by industrial consortia, also came from the Karlsruhe research laboratory. The Karlsruhe project management carried out the first consultations with the firms which were to participate in the work, and provided the economic and technical information for the formal decision-making procedure.

The decision to terminate industrial work on the steam-cooled fast breeder was actively sought by the proponents of the sodium-cooled fast breeder in the Karlsruhe laboratory. Nevertheless, the ministry took the lead in the critical phase of the decision-making process. The die was cast in direct talks between the reactor section of the ministry and the reactor department of AEG, the firm leading the consortium for the steam-cooled fast breeder. Confronted by the ministry with the alternative either to invest some of its own funds or to agree to termination of steam-cooled fast breeder activities, AEG decided that the continuation of the work was not worth a contribution of its own funds.

The decision to construct the SNR-300 was predetermined by the institutional momentum of the program, therefore being equivalent to lack of a decision to change the direction of the program. If the teams of Interatom and Belgonucléaire were to be kept together, construction of the SNR-300 had to start within a short time. Contrary to the 1969 decision on the steam-cooled fast breeder, the sodium-cooled fast breeder was not an isolated national development, and a reversal of policy would have occasioned worldwide attention. Therefore, the ministry preferred to let the program go on in its previously determined direction.

Over the years there were significant changes in relations between the ministry and the Karlsruhe laboratory.[12] Up to the mid-1960s there was close contact between senior scientists at Karlsruhe and senior civil servants of the ministry. The reactor section of the ministry did not play an important role in decision making. It could happen that minor decisions taken by the reactor section were altered by an intervention from Karlsruhe at higher levels.

The 1966 decision to start work on prototype design marks a change in this relationship, since the reactor section then began to use the participation of industrial firms as a means to strengthen its control over Karlsruhe. Whereas Karlsruhe wanted to keep its leadership in the program and suggested that the laboratory give the design contracts to the firms, the reactor section insisted that the ministry give the design contracts and that Karlsruhe's activities by subordinated to the work of the industrial consortia. In the 1969 decision on the steam-cooled fast breeder the reactor section was in the main current of decision making. By that time, it had its own view on matters of reactor development and was relatively independent in its judgments from Karlsruhe or from its advisory bodies. This is documented not only by the decision on the steam-cooled fast breeder but also in decisions in other projects at that time, including the termination of the thermionic reactor and the cancellation of the high-temperature gas-cooled prototype plant Kernkraftwerk Schleswig-Holstein. The emerging strength of the ministry also had repercussions on the position of the advisory bodies. Unlike its predecessors, the nuclear program in 1968–1972 was drafted by the ministry, not by the Atom Commission. In 1971 the ministry eventually dismantled this time-honored advisory apparatus.

The influence of the Karlsruhe laboratory decreased considerably in the course of the negotiations on the SNR-300 plant, when its previous cost estimates became untenable. The main center of decision making was then between the ministry and the utilities as the future owner and operator of the SNR-300. Given the precedent of the West German light-water demonstration plants, the construction of the SNR-300 would have been hardly justifiable for the ministry without a financial contribution from the utilities. Among the ministry's clientele, the utilities now become the most influential. Nevertheless, the utilities used their influence mainly to reduce economic and technical risks, while the ministry remained the dominant actor in policymaking.

A dominant influence of government organizations on a government-financed R&D program is not provided for by the theory of Galbraith, but it fits well the view of Eads and Nelson (see chapter 1). Is it an idiosyncrasy of the West German institutional setup, or can this finding be generalized?

The list of large government-sponsored programs, which were initiated and pushed by government organizations, can easily be extended. The information provided in this study on the origins of the collaboration between West Germany, Belgium, and the Netherlands shows that in Belgium and the Netherlands the FBR was also pushed mainly by government organizations. In these countries, the problem was also to define a function for the government research establishments. Apart from the firm Belgonucléaire, which lived on R&D contracts, the industrial firms entered the joint program later than the research laboratories, after the governments had committed themselves to financial support.

The British project for a supersonic civilian aircraft, as another case, also originated in a government organization.[13] The first generation of nuclear power plants in the United States, Britain, and France was an offspring of government programs for nuclear weapons, and most of the government-financed R&D in this field was performed by the same government organizations as those responsible for nuclear weapons technology. The FBR programs in the United States, Britain, and France have emerged from the same government organizations, and the main part of R&D on fast breeders is still performed by them.[14]

Although Galbraith's theory does not provide for big government-sponsored programs initiated by government organizations, only a small though essential modification of his concept of the technostructure is required to accommodate it to the findings of this study. Galbraith speaks of technostructure regarding big corporations, not government laboratories. The technostructure comprises scientists and engineers who, because of their scientific and technological expertise, have influence on the corporate decision-making process. The objectives guiding their behavior are prevention of interference from shareholders and banks, corporate growth, and technological virtuosity. If one assumes that a technostructure of similar kind with similar objectives exists in government organizations performing or managing R&D, it is no longer surprising that large-scale government projects originate from and are pushed by government organizations. The case of the West German fast breeder suggests that influence from government organizations may be at least as important as influence from nongovernment organizations. Government organizations have similar goals and interests as the technostructure: survival, autonomy, organizational growth, and technological virtuosity. These may be important factors in policymaking.

Galbraith assumes that, apart from direct contacts in the policymaking process, image building plays an important role in the relationship between the industrial system and the state. With an appropriate image, the industrial system can ensure a favorable reaction to its needs. This view is supported by the case of the West German fast breeder, if the notion of the technostructure is amended to include government organizations. There was indeed considerable image building: the program was closely linked to the propagation of big science in West Germany, and it was heralded as the beginning of a new era in the relationship of government, science, and industry.[15] But this image was promoted by Karlsruhe scientists and ministry officials. The absence of industrial speakers supporting this image is conspicuous.

The lack of parliamentary control features in many analyses of science and technology policy.[16] While other authors assume that the absence of central control opens a government agency to the conflicting interests of industrial firms, it appears from this study that the lack of central political control widens the scope for government organizations to follow their own interests and goals. Conflicts among a government agency's clients increase rather than decrease the agency's autonomy. The agency can play one group of clients off against another, for instance, manufacturing industry against a government laboratory or utilities against manufacturers. As a matter of fact, the ministry used, in the 1965–1966 decision on prototype design, the participation of manufacturing firms as a means of increasing its control over the Karlsruhe laboratory; and in the decision to construct the SNR-300 it used the participation of the utilities to increase its control over manufacturing industry. In this way, an organizational setup which seemingly reduces an agency's influence can actually increase it.

There is a general conflict between the producer of a technology and its user. The case of the SNR-300 plant suggests that a government agency can utilize this conflict in order to improve its decision-making capability, wherever there are competent users. Early participation of user industry will improve a project, since attention to user needs is an essential factor in technological innovation. Channeling subsidies via the user will improve the contractual terms of government-financed projects, and will increase the amount of sound information available to the agency.

The finding that a government agency has a relative autonomy toward its industrial clientele is contrary to widely accepted views on government-industry relations. It may therefore be important to stress that the considerations here are confined to government support to R&D. In other sectors of government, this may or may not be different. As to government-financed R&D, however, the above considerations suggest that the finding of this study may be capable of generalization.

Stating a lack of central political control implies a negative value judgment about the relative autonomy of government organizations involved in a government-financed R&D program toward those central political institutions that are in charge of overseeing and controlling the activities of individual departments and their dependent organizations. To put this into perspective, one must stress that in a realistic system of government there is a balance of autonomy and control. Autonomy is therefore not necessarily something bad. This is a question of how much in relation to what function.

A program that is initiated and pushed by government organizations is inconsistent with generally accepted views on government bureaucracies. Many think that government bureaucracies should have an active rather than reactive policy but in fact do not or not sufficiently.[17] As far as government support to civilian technology is concerned, there is little reason to disagree as to the general desirability of an active behavior of government bureaucracies. In R&D, initiatives are fostered by such motivations as curiosity, technical fascination, the need to occupy redundant staff, or the desire for organizational growth. These motivations may be highly serviceable in opening new technological options. Without technical fascination, for example, many ideas would not survive the initial phase of R&D when technical uncertainties are high and economic prospects are difficult to assess.[18] However, this is different in the later, more costly stages of the innovation process. Here a realistic assessment is decisive, and indulgence in technical fascination may lead to economic disaster. At this stage, initiatives need thorough evaluation before they are carried out into action. The effectiveness of government support to the later, more costly stages of the innovation process is dependent on the ability and willingness of government organizations to make a realistic economic evaluation and to take it into account in their decisions. The findings of this study suggest that too much autonomy may induce government organizations to put their own organizational goals ahead of the general objective of an efficient allocation of public funds.

Economic Assessments

In all decisions analyzed in this study, industrial actors had more realistic views on the FBR's economic potential than government actors. At the start of the program in 1960, the Karlsruhe project management expected that breeders would be used after 1970. This date was taken from the schedules of foreign programs, not from considerations of economic need. In the advisory bodies, however, the more skeptical views of industry prevailed and led to a recommendation to consider the possibility of terminating the program after three years. There was at that time an ample supply of cheap energy, especially fossil fuels. Nuclear power was not yet competitive with fossil fuels. Opinions about when it would become competitive varied widely, though it was clear that LWRs would be used before fast breeders. The acceleration and expansion of the Karlsruhe program in the course of the negotiations with Euratom in 1961–1963 were done without any economic evaluation. In fact, this was contrary to the earlier cool appraisal by the advisory bodies.

In 1965–1966, the precipitate start of design for two 300-MWe prototypes, one steam-cooled and one sodium-cooled, was urged by the Karlsruhe laboratory on the grounds that the Americans were going to start a sales attack in the 1970s and that estimates of electricity costs showed a clear advantage of the fast breeder over the LWR. These justifications were, however, not discussed explicitly by the advisory bodies. Implicitly they were rejected, since industrial representatives in these bodies declined to contribute their firms' own funds on the ground that the economic use of fast breeders was not foreseeable.

The decision to construct the SNR-300 prototype plant was based on an economic evaluation that, in terms of man-hours spent in committee sessions, was the most extensive ever in a decision in the West German fast breeder program. This was due to the prodding by ministry officials, who took part quite actively in the consultations of the subcommittee of the Working Group, not in order to influence the direction of the subcommittee's recommendation but to see that the subcommittee worked out its analysis in detail. As a result, the document produced by the subcommittee addressed quite well all major questions. Nevertheless, the advice it presented was limited. The subcommittee relied on evidence submitted by supplier firms. Members of the advisory bodies with utility background regarded this to be an insufficient basis for decision. The most critical question, whether to build the SNR-300 plant or to seek instead collaboration in a foreign program, was handed back to the ministry.

The resulting economic assessment was biased toward the optimistic side, though it was less optimistic than official views in other countries. The decision to construct the SNR-300 was justified by the expectation that high-grade uranium reserves would be exhausted by the 1990s, and rising uranium prices would then help the LMFBR to economic competitiveness. The subcommittee saw that the depletion of high-grade uranium reserves by the 1990s was by no means certain; therefore it regarded the LMFBR as an insurance policy rather than a definite necessity.

The uncertainty about the economic need for the LMFBR was underlined by the economic analysis in chapter 7. If realistic assumptions are made about further discoveries of high-grade uranium ores and about future electricity demand, depletion of these ores may recede into the middle of the next century. In view of the high capital costs of LMFBRs and the small contribution of uranium costs to electricity costs in LWRs, very high uranium prices are required to make the LMFBR competitive. It is uncertain when these will be reached. As an insurance against long-term uranium shortages, advanced converter reactors based on proven reactor types, such as the LWR or the heavy-water reactor, are more flexible, once technologies for reprocessing uranium fuel or thorium fuel are available on a larger scale.

The West German FBR program was a response not to an urgent economic need but to a technological opportunity to be seized. Up to the present an urgent need for this technology has not been established. An urgent need may be defined in this context as the reasonable certainty that a technology will become commercially competitive within a time scale of two decades or less, and that economic advantages then will be of an order which allows the recoupment of development costs.

As a result of inadequate attention to economic need in the policymaking process, justifications for the decisions were highly volatile. The impression of an impending American sales attack, which was to justify the precipitate start of prototype design in 1966, had vanished by the time prototype design actually started. The cost estimates used as a justification in the same decision proved untenable a few years later. A good number of technical arguments used in the 1969 decision to cancel the steam-cooled fast breeder became invalid through further development activities within two years, though in this case the main justification referring to the international situation remained valid. Prospects for collaboration with the British were used in justifying the construction of the SNR-300 plant, but had faded away by the time construction actually started.

An imagery of strong competition from foreign programs has the same function as an economic evaluation and thus can complement or substitute for the latter in policymaking. Arguments in this vein were proposed by the Karlsruhe laboratory, especially in the 1965–1966 decision for a precipitate start of prototype design and in the 1971–1972 decision to start the 300-MWe sodium-cooled prototype. But they were not shared by manufacturing industry, utilities, and, for the SNR-300 decision, the ministry. In the 1965–1966 decision, the impression of an impending American sales attack was in contradiction to the view of industry representatives that the economic use of fast breeders was not foreseeable. In the 1971–1972 decision, an assertion of international competition was not consistent with an economic assessment which viewed the fast breeder as a long-term insurance policy against the possibility of depleting high-grade uranium ores.

It should be noted that the competition was with foreign government programs, not with foreign industrial firms financing the development from their own funds. Given the absence of an evaluation of the economic need for fast breeders in the early years of the West German program, given further the view of industrial representatives in the advisory bodies that an economic use of fast breeders was not foreseeable, one should speak of a technological race between government programs rather than of international competition.

Arguments referring to supposed international competition lend themselves well to the exigencies of bureaucratic and political decision making. The comparison with other countries is a fruitful source of questions for critics. Doing what others do is therefore much easier to justify than deviating from what others do. This would suggest that there is a tendency in government R&D to imitate others, whereas economic wisdom indicates that the premium is for specialization, that is, "to do what others are not doing."[19]

Inadequate attention to economic need is in accordance with the view of Galbraith as well as with that of Eads and Nelson. The difference between the two views lies in the assessment of the positive side effects of government-financed R&D in terms of extending the corporation's capability for planning and underwriting of industrial technology. Whereas Galbraith deems these side effects to be crucial for the functioning of the modern industrial system, Eads and Nelson regard them as being too small to justify government-supported programs. The self-image of the Karlsruhe scientists corresponds to Galbraith's view. Although the scientists were highly confident in a direct economic benefit of the FBR program in terms of low electricity costs, indirect benefits also ranked very high in their thinking.[20] For them the goal of the fast breeder program was twofold: to develop the fast breeder and to learn how to do big science. No matter what specific technology was being developed, any big science project was considered a general stimulation of industry.

The reality of the West German FBR program, however, corresponds to the view of Eads and Nelson. There were no technical spillovers to the development of LWRs, as the two technologies were too different for the LWR to benefit from R&D on FBRs. In terms of manpower and employment, there are indications of negative side effects. The rapid expansion of the Karlsruhe FBR activities in the early 1960s drew talented university graduates away from industrial activities on the LWR, because Karlsruhe as a government laboratory provided safer employment and better pay.[21] For the same reason, there was very little mobility of manpower from the Karls-ruhe laboratory to industry.

There are indications that the employment provided by the FBR program was not much needed by the larger firms such as Siemens, AEG, MAN, or GHH. When Siemens decided in 1969 to buy Interatom and to transfer its FBR activities to this company, it was in a position to let the members of its FBR team choose whether they wanted to move to Interatom or do other work with Siemens. About two-thirds of the team remained with Siemens. The activities of MAN and GHH on components for the steam-cooled fast breeder were terminated without problems after this type of reactor was abandoned. The completion and recording of some experimental work may have lasted until 1971, but then the teams were dissolved. For AEG, the FBR program did provide continued employment of the HDR reactor team. However, when the steam-cooled FBR was dropped, AEG's team was eased out of the program through some smaller contracts on the sodium-cooled type. After these contracts expired, the members of the team were employed in other company activities.

Thus Interatom is the only West German firm for which the participation in the FBR program was of vital importance, since this program accounted for the major part of the firm's activities. If a firm, however, lives nearly totally on government R&D, Galbraithian assumptions on the positive side effects of government-financed programs do not apply, as there are virtually no market-oriented activities that need underwriting. The frequent changes in the ownership of Interatom suggests that the firm's government-financed fast breeder activities were hardly crucial to the corporate strategy of its parent companies. Kraftwerk Union's attempt in 1969 to trade independence in fast breeder development against a major share of the French market in LWRs points in the same direction.

Observations on the policymaking process confirm that, apart from firms living on government R&D, the indirect benefits of government-financed R&D to industrial firms are marginal. If government grants were of crucial importance, one would expect controversies in the advisory bodies among the proponents of different technologies. Firms would try to increase their share of the cake to the disadvantage of others. The proponents of competing technologies would subject the claims of others to scrutiny and criticism. And conflicting interests would be settled in a political bargaining process. But such processes could not be noted in the consultations of the advisory bodies, nor were there indications that they took place outside the formal decision-making procedure. On the contrary, the firms showed a good deal of readiness to share the cake with others: AEG with GHH and MAN, Siemens with Interatom, and Interatom later with Belgonucleaire and Neratoom. This is not to say that the firms liked to share the cake with others. In fact, some interviewees indicated that the firms would have rather liked to avoid the inevitable friction entailed in collaboration. The point is, however, that they did not make it an issue in the decision-making process.

The findings of this study suggest some generalizations as to the limits of economic evaluations in government-financed R&D. As summarized in chapter 7, manufacturer and user firms are the main source of expertise. Government departments and laboratories are not well equipped for economic assessments because of their inexperience in commercial business; this is also the case for firms without commercial experience. The importance of experience for realistic economic and technical assessments is demonstrated by the fact that in West Germany only manufacturing firms with prior experience in conventional power plants achieved commercial sales of nuclear plants.

Firms have little incentive to contribute economic and technical information to policymaking, as this information becomes accessible to competitors or customers and thus may have unfavorable repercussions on their business. More important, any serious discussion of the economic prospects of a new technology implies a conflict with assessments put forward by other organizations and interferes with interests of other organizations. There are little rewards for engaging in such conflicts. In the policymaking process, the behavior of government and industrial actors is therefore generally one of conflict avoidance or nonagression.[22] Even organizations that are not direct recipients of government funds have little incentive to be critical about economic assessments advanced by others.

In the policymaking process on government support for R&D, a close examination of economic prospects will be undertaken only to the extent urged by the government. If the government is really interested in these prospects, it must discern the critical or key questions and must ask them in a detailed way. In the later decisions analyzed in this study, the termination of the steam-cooled fast breeder and the construction of the SNR-300, the ministry did this. To make sure that every member of the advisory committee expressed his view, the ministry required them to answer its questions individually and in written form. But as the decision on the SNR-300 shows, a government may get limited advice even if it asks the critical questions in detail.

Other mechanisms may be available to governments to procure advice, which have not been used in the West German fast breeder program. For example, one might commission organizations not directly involved in a program such as engineering consultants or university institutes for an economic analysis. Because of disappointing experiences in other projects, the ministry did not seek the advice of engineering consultants in the fast breeder program.[23] This experience confirms that even actors not directly involved in a program have stronger incentives for conflict avoidance than for criticism. There seems to be hardly any organization which both commands sound technical and economic information and has incentives for critical evaluation. This does not mean, however, that studies commissioned to engineering consultants or university institutes may not be worthwhile. A government department may to some extent get round the conflict-avoiding behavior of its clients by posing specific questions to different organizations. Nevertheless, there are limitations resulting from the fact that the basic sources of economic expertise are, in any area of technology, the firms that produce or use the technology. All other organizations depend on them for original information.

Another mechanism might be publication of recommendations and justifications of advisory bodies, perhaps including minority votes of dissenting members. This would subject such advice to a public debate. There are precedents for this kind of advisory body in other countries, for instance the Royal Commission on Environmental Pollution in Great Britain or the Ranger Uranium Environmental Inquiry in Australia.[24] It may well be worth the effort to apply this procedure to decisions in government R&D policy. But it also has limitations. The quality of advice will depend on the extent to which members of a commission or a committee have access to sources of original information. The requirement to publish the recommendations of advisory committees is unlikely to increase incentives for committee members to be critical.

Altogether, the findings of this study suggest that governments will find it difficult to get good advice. For organizations receiving government funds, poor market prospects will not suffice to reject the funds. Other organizations not directly involved have no incentives to be critical. If a government agency really wants to find out about the economic prospects of a project, it has to invest a considerable analytical and political effort. This kind of effort is not always compatible with an agency's organizational objectives, and the agency is then likely to refrain from it.

However, incentives and reward are created for critical assessments if firms are urged to make a contribution from their own funds. When it comes to plant construction, commercial-type technical guarantees can fulfill a similar function. Only these two mechanisms are likely to trigger an extensive review of economic and technical factors by the participating firms themselves. A firm's representative can take part in the consultations and recommendations of an advisory committee without prior inside analysis in his firm. A contribution of the firm's funds, however, involves a decision-making process in the firm itself and will finally be decided by the top management. Also a commercial-type guarantee will be signed only after the technical risks are assessed in detail and the top management has been satisfied that the risks are acceptable to the firm. Equally important, a requirement that firms contribute their own funds will not only induce firms to make thorough economic assessments, but will make them willing to communicate their assessments to policymakers.

Development Strategy

The existence of a time-cost trade-off was confirmed in this study, though it was not always possible to give exact figures for the cost saving that could have been achieved if the program had proceeded in a sequential and incremental manner and at a slower pace. The late start of the West German program can be regarded as an instance of time-cost trade-off, though unintentional, as it was due to the political circumstances of postwar Germany rather than to deliberate choice. As the West German program started about fifteen years after that of the United States, and about ten years after that of Great Britain, a lot of relevant know-how was available simply by monitoring what the others had done, practically without cost. Technical developments such as the transition from metallic to oxide fuel had rendered a good deal of the earlier development efforts obsolete. The actual lag of the West German program behind foreign programs was thus much smaller than the year-span would suggest.

The West German program differed from the programs in the United States, the Soviet Union, Britain, and France in that it did not follow the usual sequence of having first a small test reactor and thereafter a prototype reactor of several hundred electrical megawatts. Instead of a sequential strategy, there was a good deal of overlap between the test reactor KNK and the prototype SNR-300. This overlap was accentuated by the fact that prototype design was rushed at the quickest possible pace, while no hurry was felt at all in providing a domestic fast test reactor. With hindsight, it can be clearly seen that doing things in the usual sequence would have been cheaper. Because the test reactor was not provided in a timely manner, there was a lot of tinkering with thermal irradiations, which were later dismissed as unsystematic and of limited applicability. The relevant test irradiations were done in foreign test reactors, such as the British DFR and the French Rapsodie. The considerable expenditure for these irradiations and for those in the Belgian BR2 reactor could have been saved if the program had proceeded in a sequential manner.

A sequential strategy would have also saved some expenditure in the design work on the steam-cooled prototype. If the schedule had provided for installing a fast or a fast-thermal core in the HDR reactor before starting the design for a 300-MWe prototype, the situation of the steam-cooled FBR in 1969 would not have been different from that which finally led to the termination of the steam-cooled fast breeder. There would have been the same amount of information on which to base this decision, but with

DM30 million less expenditure. A delay of construction of the SNR-300 by only a few more years would have resulted in some savings in operation and construction costs. It would have put the plant on a sounder technological basis and would have reduced the risk of technical failures.

The most important instance of a time-cost trade-off concerns the timing of the FBR program relative to large-scale use of plutonium in proven types of reactors. Large-scale use of plutonium entails many uncertainties with regard to technology, economics, licensing, and social and political implications. The basic technology of the plutonium fuel cycle is relatively the same, whether the plutonium is used in fast breeders or in thermal reactors. Although the FBR fuel cycle has some cost penalties due to higher concentrations of fissile material, higher burnup, and more intense radiation in spent fuel, its technology builds on that of thermal plutonium recycle. An FBR program better tuned to the development of the plutonium fuel cycle for proven types of plants would have implied a delay for fast breeders, but it would have greatly reduced the fuel-cycle cost of the SNR-300 and would have put decisions on a sounder basis with regard to the economics and technology of the plutonium fuel cycle. Thus much more information would have been created at the same cost, or the same amount of information at a lower cost.

There are a variety of reasons why time-cost trade-offs were disregarded in the policymaking. Unrealistic economic assessments by the Karlsruhe laboratory led to far too optimistic expectations for the market introduction of fast breeders. The program schedule was set in the early 1960s with respect to the schedules of foreign programs rather than to economic need. A misplaced sense of international competition and the politics of European collaboration caused unnecessary accelerations. Some of these factors, combined with a mutual optimization of organizational interest between Karlsruhe and Interatom, were also responsible for the overlap of the KNK and SNR-300 projects. Finally, but perhaps most important, there was institutional momentum. The proposed date for FBR commercialization slipped by one or two decades when the decision on the construction of SNR-300 was to be taken. There was a case for delaying construction of the SNR-300 as long as possible in order to reduce the costs of bridging the period until commercialization. However, there was little flexibility in the schedule, if the industrial teams were to be kept together.

In general, government actors were less sensitive to technical and economic uncertainties than industrial actors. The assessment of technical risks in the policymaking shows the same dynamics as economic evaluation. Industrial actors have little incentive for pointing out technical uncertainties, unless they are asked to commit themselves either by a commercial-type guarantee or by a contribution to project costs from their own funds. As long as only government money is at stake, firms are prepared to take technical risks they would not take in a self-financed program. The limits of advice described above with regard to economic evaluation extend also to technical evaluation.

The role of conflict avoidance is illustrated by the project committee's recommendations in the decision to terminate the steam-cooled fast breeder. Because of the difficulties in transforming the HDR reactor into a steam-cooled fast test reactor, AEG suggested to the project committee in April 1968 that it terminate design work for the 300-MWe prototype and design instead a new small steam-cooled fast test reactor. The project committee approved the proposal. In September 1968 AEG presented to the project committee a more detailed version of its proposal; the project committee confirmed its approval. In January 1969, after AEG had given its consent to a termination of the steam-cooled fast breeder, the ministry presented to the project committee its intention for a policy reversal. The project committee agreed. Between the latter two sessions, there had been no additional technical or economic information. The only change was that the ministry had made known its determination for a reversal of policy, and that AEG had agreed to it. After all, the main argument for terminating the steam-cooled fast breeder reactor was true three years earlier, when design of a 300-MWe prototype started. In 1966 West Germany was isolated in the international context. No steam-cooled fast breeder was under construction or in operation anywhere in the world, and the design work done in foreign countries was only of a preliminary kind.

The utilities were the only actors to press for some delay in order to reduce technical risks. In the decision to construct the SNR-300 the Working Group of the German Atom Commission paid insufficient attention to technical and economic risks, recommending construction start as quickly as possible. It needed the pressure from utilities to reduce the risks entailed in a premature construction start through a further delay of one year. No such pressure was applied in the earlier decisions as long as they did not have a financial stake in the venture.

The need for a redesign of the SNR-300 plant with regard to user needs shows that an advisory committee like the Fast Breeder Project Committee cannot adequately supervise the design activities for a prototype plant. Advisory committees meeting a few days per year are not in a position to analyze and to determine the design objectives of a prototype plant in sufficient detail. However, this job can be done directly between the utilities and the supplier, if the design contracts are given by the utilities rather than directly by the government, and if a financial contribution from the utilities provides incentives for them to look thoroughly at the designs and to extract guarantees from the supplier. This lesson has been applied to the SNR-2 design.

If the behavior of the actors in the West German fast breeder program can be derived from rather general assumptions, there is little reason to believe that it was exceptional or idiosyncratic. As far as conflict avoidance is concerned, enough has been said earlier to suggest that it is not unique to actors in the West German FBR program. Institutional momentum, as one of the main factors militating against a realization of time-cost trade-offs in government-financed R&D, appears also to be a general phenomenon. There is a minimum level of effort in any R&D project that is defined by the requirement of keeping a team together. If a team is dissolved, a good deal of knowledge is destroyed or lost. A new start with a new team requires time and cost until the previous level of expertise is reached. The minimum level of effort required for continuous employment of a team may be very low if the organization has a sufficient number of different projects on which team members can exercise their skills. However, if the organization works primarily on only one project, it faces an unpleasant alternative: either terminate the project and dissolve the team or continue at a level of effort much higher than the optimum for time-cost or time-risk trade-offs. When a prototype plant is constructed, the design teams finish their job long before construction is completed and operation starts. If there are no other projects to work on, the teams therefore have to begin work on the next, larger plant, before the prior plant yields significant operation experience.

Few people would deny that employment considerations have a right of their own. This suggests that such considerations be made before unpleasant alternatives arise. Policy decisions should be made so as to foreclose their possibility. One way to achieve this may be placing a ceiling for the proportion of its staff a firm may employ for one government-financed project. Similar considerations would apply, of course, to government laboratories.

Scale of Costs

In the case of the West German fast breeder, the financial contribution of the participating firms was not limited by the firms' financial capabilities but by the fast breeder's poor economic prospect. In the 1960s, reactor vendors and utilities refused to contribute their own funds because an economic use of fast breeders was not foreseeable. In the 1970s, industry did contribute some funds. But most of this contribution came from the utilities, and the major part of the utilities' contribution was to the commercial component in prototype construction costs. The financial contribution by manufacturing industry is still not much more than nominal.

As shown in chapter 7, West German reactor manufacturers and utilities would be able to finance a 1,300 to 1,500 MWe demonstration plant from their own funds, provided the technology lives up to the promises made by its proponents. Design and R&D would take less than 2 percent of Siemens's self-financed R&D. If construction of the demonstration plant would cost 50 percent more than an LWR, the noncommercial component that would not be recoverable from electricity sales would require a yearly expenditure that would take a similar proportion of RWE's sales income as the utility's VAK experimental boiling-water reactor. This calculation suggests that industry does have the financial capability to develop an FBR up to commercial introduction, if it can draw on a government-financed program for exploratory R&D and if risks in the operation of prototype and demonstration plants are shared by government.

Exploratory development is defined as testing broad attributes of new product and process designs (see chapter 1). In the development of nuclear reactors, this definition is not sharp and can be interpreted in different ways. The above calculations have implicitly assumed that exploratory development may include the construction and operation of a 300-MWe demonstration plant. This is rather an extensive interpretation. Let us therefore speculate whether a more narrow interpretation would be feasible with regard to the financial capabilities of reactor vendors and utilities, assuming a clear economic need for the FBR assured an adequate rate of return.

In a narrow interpretation, we may understand exploratory development to include work on reactor physics and coolant technology up to the construction of test rigs for cooling components, and the construction of a test reactor for fuel irradiation. The test reactor might have an output of 100 to 400 thermal megawatts, that is, smaller than the SNR-300 plant (770 MWt) but larger than the KNK II plant (58 MWt). If the Karlsruhe activities in 1960–1982 and the costs of the KNK reactor are taken as a rough guideline, the costs of a program for exploratory development in fast breeder technology could be estimated at DM2,000 million to DM2,500 million. The exact costs of such a program would be determined by the amount of exchange of know-how with foreign programs.

The construction of prototype and demonstration plants is the largest cost item in the development of power reactors. A good deal of these costs can be recouped from sales of electricity, and this commercial part of the construction costs is larger the nearer a plant comes to competitive electricity costs. If protected by government risk sharing, the commercial part of the construction costs can be financed by utilities. The uncommercial part of the construction costs would have to be paid for by the manufacturer or the user, or both. If the commercial prospects of fast breeders were such as to justify full-scale development, fuel-cycle costs would have to be of the same order of magnitude as light-water fuel-cycle costs. Capital costs of a 300-MWe fast breeder demonstration plant would have to be not more than 250 percent of LWR capital costs. The noncommercial part of the investment costs then would amount to 150 percent of LWR costs, that is, in 1979 prices, DM900 million. A large demonstration plant in the range of 1,300 to 1,500 MWe would have nearly competitive electricity costs. In addition to the noncommercial part in the costs of prototype construction, industry would have to pay for design work and engineering. If industry could draw upon a government-financed exploratory development program, it would have to put up about DM1,500 million to DM2,000 million for developing FBRs up to the first commercial order. This sum would have to be financed by manufacturers and utilities together. The amount is certainly not so high as to put it beyond industry's financial capability. Distributed over a period of fifteen years, it would take about 4 to 6 percent of Siemens's R&D budget.[25] Given the fact that some of these costs could be covered by the utilities, given further the possibility that a manufacturing firm could join with others for a specific development program, even development costs two or three times higher than those estimated above would not transcend industry's capabilities, provided the firms were assured an adequate rate of return by a clear economic need for the fast breeder.

If the scale of costs does not provide an argument for direct government subsidies to industrial R&D in the case of the fast breeder, it seems highly likely that this argument can be rejected in general, since the fast breeder is the biggest government-financed civilian R&D program ever in the Federal Republic of Germany and in other countries. It should be remembered, however, that the electrical industry is highly concentrated and that the firms in this industry are quite large. In less concentrated industries or in countries with technically less sophisticated firms, it may be possible that firms cannot or do not want to afford programs of a scale as represented by fast breeder development. But thresholds of R&D are likely to be associated with thresholds in production and marketing. Direct subsidies to R&D would in that case not be a sufficient way of government intervention.

If the cost of even large R&D projects is within industry's financial reach, it follows that firms have little motivation to seek government support, as long as they have confidence in the good economic prospects of a technology. This might make economic and technical information accessible to competitors. It opens a project to the influence of government organizations, thus burdens it with additional uncertainties. Finally, government support poses the question of sharing profits with the government. These considerations also imply that firms offer for government support only projects with poor or uncertain economic prospects or with high technical uncertainties. They accept government support for an essential fraction of their activities only if they cannot assure their survival through commercial business.

International Collaboration

In the West German fast breeder program, international collaboration has occurred mainly in three forms: collaboration with an international government agency (Euratom); forming a joint program with other countries (Belgium and the Netherlands); and exchange of know-how with foreign programs.

The case of the West German fast breeder lends support to the view that Euratom was an inappropriate instrument for both fostering technological innovation and promoting political integration.[26] First of all, the scale of costs in developing nuclear reactors is not such as to put this technology beyond the capability of individual countries of the size of West Germany. LWRs were developed in West Germany with very little government money. Fast breeders were not beyond the national capability.[27]

Second, as there was practically no role for Euratom in funding R&D on the first generation of nuclear plants, the agency spent its considerable R&D funds on what was called the second generation or advanced types of nuclear reactors. By definition, these were technologies for which an economic need was remote and, by implication, uncertain. This is substantiated by the fast breeder program. At a time when West German advisory bodies considered the possibility of terminating the program after a preliminary effort of three years, Euratom was pushing fast breeders and gave this technology the highest priority.[28] This is consistent with Pavitt's view that international agencies are left to support projects of a low national priority.[29] Moreover, this study shows that this type of international collaboration can have a negative effect on national R&D priorities. The politicking with Euratom pushed the West German fast breeder program from a low priority suddenly to the highest priority among all types of nuclear plants, and this at a time when industrial activities were concentrating on the LWR.

Third, if an international agency is endowed with its own R&D budget, decisions on allocations to specific projects are bound to create political frictions, especially regarding a fair return. Instead of a contribution to political integration, such agencies tend therefore to be a continuous source of quarrels. The case of the fast breeder is a good illustration.

These conclusions do not deny in general that international collaboration in civilian R&D may be desirable for political or economic objectives. Neither do they imply that Euratom had no positive achievements. Euratom did bring the scientists from different national programs together. The present structure of European collaboration in FBRs owes a good deal to the early contacts provided by Euratom. But if international collaboration in R&D is to make a contribution to political aims of integration or if it is to help nations to collaborate in an R&D program to their mutual economic benefit, other ways of finance appear advisable that avoid the problem of a fair return. Instead of deciding first on a joint R&D budget and in a second step on the projects to be financed, a better strategy seems to agree on financial commitments for each specific project. The SEFOR project or the projects of the International Energy Agency may be examples in this direction.

With regard to the collaboration of West Germany with Belgium and the Netherlands, this study found that political motives dominated technical and economic factors. There was no financial or technical need for this collaboration, as West Germany had the financial and technical capability to do the program on its own. In view of some work duplication, the financial benefits were less than the financial contributions by Belgium and the Netherlands suggest. The benefits were rather on the political and organizational levels. The collaboration was useful for West Germany in allaying foreign suspicions about potential military implications of the use of plutonium, and in safeguarding West Germany against potential discriminatory applications of the nonproliferation treaty. On the organizational level, the collaboration was welcome as a confirmation of the government's fast breeder policy, and as an addition to the reputation of the Karlsruhe laboratory.

For Belgium and the Netherlands, collaboration with a major country was a precondition for an engagement in fast breeder development. West Germany was preferred as a partner to France, since it offered more generous terms for the collaboration. The initiative for the collaboration came clearly from the government laboratories of the three countries. The fast breeder program was attractive to the Belgian and Dutch laboratories, since it provided a meaningful function for the longer term. Incentives were thus very similar to those in the Karlsruhe laboratory.

Possibilities of broadening the international collaboration were discussed when the decision on the construction of the SNR-300 plant was to be taken. Analysis of these discussions shows that participants in government-financed programs have no incentive to share their work with foreign institutions, as long as the government is prepared to provide the funds for an independent effort. It seems, therefore, that a general distinction is to be made between those forms of international collaboration which involve sharing the work, and those in which none of the partners has to reduce its own activities. Any collaboration between two organizations is a possible source of friction and involves costs in terms of an increased effort for coordination. But this is acceptable if offset by benefits such as an increase of reputation or political support. Certain forms of collaboration such as exchanging information and know-how or accepting the participation of foreign partners in a junior role are therefore relatively unproblematic. Real sharing of work especially in a junior role, will be accepted only if there are no alternatives. This was the case for the collaboration of Belgium and the

Netherlands with West Germany, since an independent effort was beyond the resources of these countries.

The attitudes toward the different forms of international collaboration found in this study do not appear to be restricted to the case of the West German fast breeder. Organizations performing government-financed R&D seem to perceive the benefits of collaboration in terms of their own organizational interests rather than in terms of the overall cost-effectiveness of a developing program. By implication, this would also apply to government agencies financing such programs. In practical terms, these reflections would suggest that there is less international collaboration in government-financed projects than would be desirable in terms of cost-effectiveness, and that the cooperation that does take place is often inefficient.

International collaboration is often criticized as a source of inflexibility which renders a reversal of policy difficult or even impossible. In the case of the West German fast breeder, this kind of implicit obligation did not emerge until recently, when it was said that the terms of international collaboration would make the termination of the SNR-300 plant more costly than its completion. The collaboration with the French, agreed in 1976, may lend itself to arguments of a similar vein. Without information on the legal and political details, it will generally be difficult to decide in which cases such an obligation is real and in which it is just a convenient excuse.

Decentralized Decision Making

The evidence presented in this study supports criticism of government support to full-scale development of civilian technology. The West German fast breeder program was initiated and promoted by government organizations as a technological opportunity to be seized, without a clear economic need having been established. The costs of the program were not such as to put this technology beyond the capacity of business enterprises, if these could have drawn on government-financed basic research and exploratory development. The commitment by industry of its own funds was limited not by financial capability but by the poor commercial prospects of fast breeders. The government agency did not command the economic and technical expertise necessary to prevent its commitment to a commercially poor project. Instead there were, especially in the initial phase of the program up to 1966, organizational interests which worked against a critical examination of economic need and technical risk. In the latter phase, the agency became more realistic in its judgment. By then, however, the program had acquired an institutional momentum which made it difficult to change its direction or schedule.

The case of the West German fast breeder does not appear exceptional.

There are a number of similar cases in other countries. A look at the incentives and motivations of the actors involved suggests that inadequacies in the decisions have not been the coincidental results of individual or organizational idiosyncrasies, but reflect more general factors. There are limits to the capability of government agencies and laboratories to make sound economic and technical assessments of new civilian technology, as their expertise is dependent upon that of the producer and user industries. There are limits to incentives and rewards for industry to make its assessments known, and for policymakers to take these assessments into account.

The findings of this study suggest that the decentralized decision-making framework of a capitalist economy entails incentives and rewards for firms to make realistic assessments of new civilian technology, while in government decision making on civilian technology these incentives and rewards are less strong.[30] This implies that in government-financed R&D there is a general tendency to more optimism than in industry-financed R&D, with regard to the technical and economic evaluation of new civilian technology.

If business enterprises can draw on a government-financed effort for exploratory development, the development and commercialization even of sophisticated, large, and costly technology is not beyond their financial capability, provided there is an economic need for the technology. Within the existing capitalist framework, government subsidy to commercial development (as opposed to exploratory development) therefore lacks a persuasive rationale. A decision-making strategy that makes the commercial development of new civilian technology dependent on industry's willingness to finance it with its own funds, seems a better use of societal resources than government subsidy of commercial development. Where market imperfections call for government intervention, exploratory development seems preferable to commercial development as target for government support. In the case of the FBR, this analysis suggests that, provided the safety and proliferation risks can be reduced to acceptable levels, government support to exploratory R&D is sufficient with respect to the social benefits claimed for this technology. There may be cases where government support of exploratory R&D may not be sufficient in order to correct market imperfections. Then government intervention on the demand side so as to assure that economic need is adequately signaled by market demand may be more effective than subsidies to commercial development.

The proposed decision strategy is no panacea. It may compensate to some extent for the lack of incentives for realism in government decision making, and may help reduce optimistic bias. By no means does it preempt other possibilities to increase incentives for realistic appraisals, for example, more open policymaking, more outside review by independent individuals, or frank exposure of errors to improve opportunities for learning.

These considerations take the existing capitalist framework as a framework of reference. They do not forward a general advocacy of a private-enterprise economy. Technological innovation is an important but not the only criterion for assessing an economic system. Other criteria may be equally important; for example, the capability of an economy to control undesired consequences of new technologies with regard to employment and environment. Finally, countries with a nationalized industry may find functional equivalents to the decentralized policymaking strategy proposed in this study. What seems important in policymaking in civilian technological innovation is not the question of private or public ownership, but incentives and rewards for realistic appraisals.

Countries with a more centralized structure of their power-plant manufacturing and electric-utility industries may create some incentives for realism by choosing an organizational framework for construction and operation of demonstration plants that, safety and environmental regulation notwithstanding, is as close as possible to usual ways of business in the power industry. Also in a centralized framework, guarantees with regard to technical performance and price provide incentives for manufacturers to make realistic economic assessment and, equally important, to make these assessments known in the policymaking process. On the side of the utilities, similar incentives may be created by requiring them to charge the full construction and operation costs of demonstration plants to customers' electricity bills.

Again these measures are no panacea. The oligopolistic competition of advanced capitalism seems to provide for incentives to be realistic for which a functional equivalent may not be easy to devise. Perhaps decentralization of decision making and some sort of competition may be able to provide such incentives also in a noncapitalist framework.

Notes

1. As for the Finance Ministry, its influence was perceptible in the decisions to construct the KNK and HDR plants, though not in the fast breeder program proper. Negotiations between the Federal Ministry for Scientific Research and the Finance Ministry delayed the construction of these plants by a few years.

2. The chairman of the Bundestag Committee for Nuclear Energy and Water Economy from 1961 to 1965, K. Bechert (SPD), expressed at various occasions his skepticism concerning the safety features of the FBR; see, for example, minutes of the committee's 20th session, 13 November 1963, p. 16 (unpublished). In 1974 K.-H. Kern (SPD) called for a termination of government support because of the cost increases in the program; see

K.-H. Kern, "Den 'Schnellen Brüter' weiter fördern? Kostenexplosion stellt wirtschaftlichen Nutzen in Frage," *Sozialdemokratischer Pressedienst Volkswirtschaft* 29 (19 September 1974):4–5. A few days later, however, Kern's colleague Flämig (SPD) moderated Kern's proposal, saying that government support should continue, but the subsidies to the SNR-2 plant should be less than of an order of DM1,000 million; see G. Flämig, "Kein Stop für 'Schnelle Brüter'. Internationale Zusammenarbeit ist zu verstärken," *Sozialdemokratischer Pressedienst Volkswirtschaft* 29 (1 October 1974):1–2.

3. Because of apprehensions about the fast breeder's safety features, some government funds relating to SNR-2 were frozen in the 1977 budget. In May 1977 the budget committee of the Bundestag demanded that no new obligations for industrial work on the SNR-2 be made until the government had answered a series of questions. For the government's reply see "Bericht des Bundesministers für Forschung und Technologie über die Entwicklung des Natriumgekuhlten Schnellbrutreaktors," mimeographed, 1 September 1977.

4. See "Neue Konzeption für das Kernkraftwerk Kalkar/Niederrhein," *Wirtschaft in Nordrhein-Westfalen* (29 September 1978).

5. "Bericht über den Stand der Arbeit und die Ergebnisse der Enquete-Kommission 'Zukünftige Kernenergie-Politik' des 8. Deutschen Bundestages," Bundestagsdrucksache 8/4341, June 1980.

6. In addition to the articles referred to in chapter 5, see "Unzulängliche Kontrolle der Grossprojekte," *Frankfurter Allgemeine Zeitung* (FAZ) (18 February 1970); "Falsche Prognosen im Projekt Schneller Brüter," *FAZ* (8 April 1970); "Zweifelhafte Wirtschaftlichkeit des Natriumbrüters," *FAZ* (29 April 1970); "Das Ende der Natriumbrüter-Illusionen," *FAZ* (6 May 1970); "Im Bundestag ist guter Rat teuer," *FAZ* (3 February 1971); "Kostenexplosion des Natriumbrüters," *FAZ* (9 June 1971); "Das Natriumbrüter-Spiel wird immer teurer," *FAZ* (8 December 1971); "Natriumbrüter: Kostenexplosion und keine Folgen?," and "Das Milliardenspiel," *FAZ* (7 February 1972); "Voraussehbare Fiaskos in der Kernenergietechnik," *FAZ* (21 August 1974); "Schluss mit dem Natriumbrüter," *FAZ* (25 September 1974).

7. "Ein FAZ-Journalist kampft gegen das Establishment," *Capital* (February 1968).

8. For a cursory historical sketch see P. Stichel, "Alternativen der Kernreaktorentwicklung," in *Wissenschaft zwischen autonomer Entwicklung und Planung—Wissenschaftliche und politische Alternativen am Beispiel der Physik,* report Wissenschaftsforschung 4, edited by G. Küppers, P. Stichel, and P. Weingart (Bielefeld: Forschungsschwerpunkt Wissenschaftsforschung, Universität Bielefeld, 1975), pp. 115–139.

9. Interessengemeinschaft gegen radioaktive Verseuchung Kalkar-

Hönnepel, letter to the Bundesminister für Forschung und Technologie, reprinted in *Dokumentation über die öffentliche Diskussion des 4. Atomprogramms der Bundesrepublik Deutschland für die Jahre 1973-1976*, edited by Der Bundesminister für Forschung und Technologie (Bonn: Bundesminister für Forschung und Technologie, 1974), pp. 558-570; "Kalkarer Furcht vor dem Schnellen Brüter," *Rheinische Post* (15 February 1973); "Stop Kalkar," *Rheinische Post* (13 September 1973); "Kirchenvorstand stolperte über den 'Schnellen Brüter'," *Rheinische Post* (25 October 1973); "Hopp, hopp, hopp—Kalkar stopp!," *FAZ* (30 September 1974).

10. "Brüterbeteiligung vom Parlament gebilligt," *Atomwirtschaft* 18 (1973):222; L.N. van den Bos, "Transnationale Aktion der Holländer am Niederrhein," *FORUM. Zeitschrift für Transnationale Politik* (February 1975):64; J. Haschke, "Die Anti-Kalkar-Welle in den Niederlanden," *Elektrizitätswirtschaft* 73 (1974):767-768.

11. "Der Beschluss des Bundesverfassungsgerichts zur Genehmigung Schneller Brüter," *Atomwirtschaft* 24 (1979):95-99.

12. Interviews with participants, February 1975, February 1976.

13. A short illuminating account is J. Barry et al., "The Concorde Conspiracy," *The Sunday Times* (8 and 15 February 1976). For more literature on Concorde see E. Braun et al., *Assessment of Technological Decisions—Case Studies* (London and Boston: Butterworths, 1979), pp. 25-30.

14. For the United States, see P. Mullenbach, *Civilian Nuclear Power: Economic Issues and Policy Formation* (New York: Twentieth Century Fund, 1963); R. Perry et al., "Development and Commercialization of the Light Water Reactor, 1946-1976," Report R-2180-NSF (Santa Monica, Calif.: Rand Corporation, 1977); for Great Britain, see M. Gowing, *Independence and Deterrence: Britain and Atomic Energy 1945-1952* (London: Macmillan, 1974); D. Burn, *The Political Economy of Nuclear Energy*, Institute of Economic Affairs Research Monograph 9 (London: Institute of Economic Affairs, 1967); idem, *Nuclear Power and the Energy Crisis* (London: Macmillan for the Trade Policy Research Centre, 1978); R. Williams, *The Nuclear Power Decisions: British Policies, 1953-78* (London: Croom Helm, 1980); for France, see R. Gilpin, *France in the Age of the Scientific State* (Princeton, N.J.: Princeton University Press, 1968), pp. 282-288; L. Scheinmann, *Atomic Energy Policy in France under the Fourth Republic* (Princeton, N.J.: Princeton University Press, 1965).

15. W. Häfele, "Neuartige Wege naturwissenschaftlich-technischer Entwicklung," in *Die Projektwissenschaften*, Forschung und Bildung, Schriftenreihe des Bundesministers für wissenschaftliche Forschung, no. 4 (Munich: Gersbach und Sohn, 1963), pp. 17-38; idem, "Die Projektwissenschaften," *Radius*, no. 3 (September 1965):3-13; W. Häfele and J. Seetzen, "Prioritäten der Grossforschung," in *Das 198. Jahrzehnt: Eine Team-*

Prognose für 1970 bis 1980, edited by C. Grossner et al. (Hamburg: Christian Wegener Verlag, 1969), pp. 407–435; W. Cartelliere, "Die Grossforschung und der Staat," in *Die Projektwissenschaften,* Forschung und Bildung, Schriftenreihe des Bundesministers für wissenschaftliche Forschung, no. 4 (Munich: Gersbach und Sohn, 1963), pp. 1–16; idem, *Die Grossforschung und der Staat,* 2 vols. (Munich: Gersbach und Sohn, 1967/ 1969); J. Seetzen, "Das Projekt Schneller Brüter: Grossforschung aus projektwissenschaftlicher und systemtechnischer Sicht," in *Vorträge gehalten anlässlich der Hessischen Hochschulwochen für staatswissenschaftliche Fortbildung 9. bis 15. October in Bad Sooden-Allendorf* (Bad Homburg vor der Höhe, Berlin, Zürich: Max Gehlen, 1967), pp. 83–108. These writings were inspired by the ideas of A. Weinberg; for a collection of his papers see *Reflections on Big Science* (Cambridge, Mass.: MIT Press, 1967).

16. As for West Germany, see J. Hirsch, *Wissenschaftlich-technischer Forstschritt und politisches System* (Frankfurt am Main: Suhrkamp, 1970); V. Ronge, *Forschungspolitik als Strukturpolitik* (Munich: Piper, 1977).

17. R. Mayntz and F.W. Scharpf, *Policy-Making in the German Federal Bureaucracy* (Amsterdam and New York: Elsevier, 1975).

18. The author thanks Robert V. Laney for making this point in a discussion.

19. K. Pavitt and S. Wald, *The Conditions for Success in Technological Innovation* (Paris: Organisation for Economic Co-operation and Development, 1971), p. 124.

20. For references see note 15.

21. Interviews with participants, September 1976, October 1976.

22. Two instances are mentioned in this study that do not conform to a behavior of nonaggression: General Electric's (U.S.) skirmish with the U.S. Atomic Energy Commission over the commission's advanced reactor strategy (chapter 4); and the recommendation of the Working Group in 1960, prodded by industrial members of the committee, to consider the possibility of terminating the Karlsruhe fast breeder program after three years (chapter 3). In both instances, conflict over a specific reactor program was embedded in a more general conflict about the division of responsibilities between government organizations and industrial firms.

23. Interviews with participants, February 1976, October 1976.

24. Royal Commission on Environmental Pollution (Great Britain), *Sixth Report: Nuclear Power and the Environment* (London: HMSO, 1976); *Ranger Uranium Environmental Inquiry: First Report* (Canberra: Australian Government Publishing Service, 1976).

25. In the business year 1978–1979, Siemens spent DM2,700 million on R&D, of which more than 90 percent was self-financed (see the company's annual report).

26. K. Pavitt, "Technology in Europe's Future," *Research Policy* 1

(1972):210–273, especially pp. 249–266. For a discussion of views on Euratom and European collaboration in general see R. Williams, *European Technology: The Politics of Collaboration* (London: Croom Helm, 1973), especially pp. 39–43, 138–151.

27. This is a point where Nau's otherwise excellent account needs correction. There was, speaking in technical and economic terms, no need for Euratom R&D funds for the development of FBRs. The stimulus for West Germany's Euratom policy, at the level of the Federal Government, was not scarcity of domestic funds but the problem of a fair return; Nau seems to present the perspective of Karlsruhe rather than that of the Federal Government. See H. Nau, *National Politics and International Technology: Nuclear Reactor Development in Western Europe* (Baltimore and London: Johns Hopkins University Press, 1974), p. 217.

28. "Euratom und die Atomindustrie," *Atomwirtschaft* 7 (1962):382–390; H.-H. Haunschild, "Die Europaische Atomgemeinschaft," in *Taschenbuch für Atomfragen 1964,* edited by W. Cartellieri (Bonn: Festland Verlag, 1964), pp. 332–353.

29. Pavitt, op. cit., note 26, p. 254.

30. The lack of incentives for realism in government decision making is also stressed by P.D. Henderson, "Two British Errors: Their Probable Size and Some Possible Lessons," *Oxford Economic Papers* 29 (1977):159–205.

Appendix A:
Chronology of Events

May 1945 End of World War II in Europe. The work of German groups on nuclear fission is terminated before a critical assembly is achieved. The allies ban activities on nuclear technology in Germany.

November 1946 A small fast research reactor, Clementine, starts operation at Los Alamos Scientific Laboratory in the United States.

May 1949 The Federal Republic of Germany is constituted as the Basic Law comes into force.

December 1951 Experimental Breeder Reactor I produces the first nuclear electricity in the world.

December 1953 U.S. President Eisenhower proposes the Atoms for Peace program in a speech to the United Nations Assembly.

October 1954 The Paris treaties are signed. West Germany renounces the manufacture of atomic weapons.

November 1954 The Physikalische Studiengesellschaft is founded by a number of industrial firms in order to prepare the finance of a reactor designed at the Max Planck Institute for Physics at Göttingen.

May 1955 The Paris treaties come into force. West Germany is permitted to work on civilian nuclear technology.

August 1955 First United Nations Conference on the Peaceful Uses of Atomic Energy, in Geneva.

October 1955 The Ministry of Atomic Questions is established. Franz Josef Strauss is named Minister for Atomic Questions.

January 1956 First session of the German Atom Commission.

July 1956 The Karlsruhe nuclear research laboratory is established with the foundation of the Kernreaktor-Bau- und Betriebs-GmbH.

October 1956 Siegfried Balke follows Strauss as Minister for Atomic Questions.

December 1956 The parliament of North-Rhine-Westphalia decides to establish a nuclear research laboratory, later located at Jülich.

January 1957 The Working Group on Nuclear Reactors of the German Atom Commission drafts the first outline of the Eltville Program.

March 1957 RWE issues a letter of intent for a 15-MWe boiling-water reactor to a consortium of Siemens AG with American Machine & Foundry Atomics Incorporated and Mitchell Engineering. The contract is later canceled.

April 1957 The AVR association of utilities gives a contract for the design of a 15-MWe high-temperature gas-cooled reactor to the consortium Brown, Boveri & Cie/Krupp.

October 1957   The Ministry for Atomic Questions is renamed Ministry for Atomic Energy and Water Economy, its responsibilities being enlarged to cover water economy.

December 1957   The Eltville Program is finalized by the German Atom Commission.

January 1958   Euratom is founded by Belgium, France, West Germany, Italy, Luxemburg, and the Netherlands.

May 1958   A joint Euratom-United States power-reactor program is agreed, envisaging the import of 1,000 MWe of American reactors to Europe.

June 1958   RWE gives an order for the 15-MWe VAK plant with a boiling-water reactor to AEG, with General Electric (United States) as subsupplier of the nuclear-steam-supply system.

July 1958   Working Group of the German Atom Commission approves the Karlsruhe laboratory application to work on advanced reactor concepts.

March 1959   First design contracts are placed by utility groups under the Eltville Program.

April 1959   Euratom issues the first invitation for proposals under the joint Euratom–United States power-reactor program.

June 1959   The Gesellschaft für Kernforschung is founded by the Federal Government and the Land Baden-Württemberg to finance an additional investment program for the Karlsruhe laboratory.

August 1959   The AVR utility group places an order for a 13-MWe plant with a high-temperature gas-cooled reactor, to be located near Jülich.

January 1960   AKS cancels plans for a 150-MWe organic-moderated reactor.

April 1960   The German Atom Commission endorses the Program for Advanced Reactors. Start of a fast breeder reactor program at the Karlsruhe research laboratory.

December 1960   Working Group of the German Atom Commission agrees to a three-year preliminary fast breeder program at Karlsruhe.

May 1961   The supervisory board of the Karlsruhe laboratory decides to increase efforts in the fast breeder program.

June 1961   The VAK plant delivers the first nuclear electricity in West Germany.

July 1961   Second invitation for proposals under the joint Euratom–United States power-reactor program.

October 1961   EBR II (25 MWe) becomes critical.

November 1961   The Ministry for Atomic Energy and Water Economy is renamed Ministry for Atomic Nuclear Energy.

June 1962   Construction start of the 50-MWe heavy-water pressure-vessel reactor MZFR, with Siemens as main contractor.

June 1962   The second five-year program of Euratom is endorsed by the

Council of Ministers. It provides for association agreements with the Karlsruhe research laboratory for fast breeder development and with the Jülich research laboratory for high-temperature reactor development.

November 1962   In a report to the president, the U.S. Atomic Energy Commission outlines a long-term program for development of fast breeder reactors. An order for a 237-MWe boiling-water demonstration plant for a site near Gundremmingen is given to AEG and General Electric (United States).

December 1962   The Ministry for Atomic Nuclear Energy is renamed Ministry for Scientific Research. Hans Lenz follows Balke as minister. The FR2 reactor at Karlsruhe begins operation at full power.

February 1963   Discussion of the Working Group with utility representatives about basic questions of a national development strategy.

May 1963   An association agreement with Euratom is notified. It entails the allocation of DM185 million for the Karlsruhe fast breeder program.

August 1963   Enrico Fermi Fast Breeder Reactor becomes critical.

December 1963   General Electric (United States) gets the order for the Oyster Creek plant, which is regarded as the first commercial order for a light-water reactor.

January 1964   The Gesellschaft für Kernforschung and the Kernreaktor-Bau- und Betriebs-GmbH are merged.

May 1964   Contracts are signed for construction of the SEFOR liquid-metal fast test reactor as a joint program of the U.S. Atomic Energy Commission, General Electric (United States), and the Karlsruhe research laboratory.

June 1964   AEG is given an order for a 240-MWe boiling-water reactor plant with fossil superheating for a site near Lingen.

September 1964   General Electric (United States) announces its confidence in offering a commercial fast breeder reactor by 1974.

March 1965   Siemens is given an order for a 283–328-MWe pressurized-water reactor demonstration plant for a site near Obrigheim.

June 1965   Construction start of the 23-MWe HDR boiling-water reactor with nuclear steam superheating.

October 1965   Gerhard Stoltenberg follows Lenz as Minister for Scientific Research.

March 1966   Start of construction of the 18-MWe-sodium-cooled zirconium-moderated prototype plant KNK.

June 1966   Construction start of the 100-MWe heavy-water-moderated gas-cooled KKN plant.

October 1966   The Enrico Fermi Fast Breeder Reactor is shut down by an accident.

November 1966 The Ministry for Scientific Research gives design contracts for a 300-MWe liquid-metal fast breeder prototype to a consortium of Siemens/Interatorm, and for a 300-MWe steam-cooled fast breeder prototype to a consortium of AEG/GHH/MAN.

December 1966   The SNEAK fast critical assembly becomes critical.

October 1967   Orders are placed for the first two commercial nuclear power plants in West Germany: the 630-MWe pressurized-water plant at Stade supplied by Siemens, and the 640-MWe boiling-water plant at Würgassen supplied by AEG.

October 1967   British scientists report new measurements on the alpha value of plutonium.

April 1968   The Fast Breeder Project Committee approves a new schedule for the steam-cooled fast breeder. Design work for a 300-MWe prototype is terminated. Industrial work is redirected to the design of a small test reactor.

July 1968   General Electric (United States) and the East Central Nuclear Group announce termination of their project for a 50-MWe steam-cooled fast breeder prototype.

February 1969   Industrial work on the steam-cooled fast breeder is terminated; activities on this reactor type in the Karlsruhe laboratory are reduced.

April 1969   AEG and Siemens merge their power-plant activities in the joint subsidiary Kraftwerk Union. Because of existing licensing agreements, the nuclear-plant activities are yet exempt from this merger.

May 1969   RWE places an order for a 1,146-MWe pressurized-water plant at Biblis.

October 1969   The Ministry for Scientific Research is renamed Ministry for Education and Science. Hans Leussink succeeds Stoltenberg as minister. The HDR reactor achieves criticality for the first time.

April 1971   The HDR reactor is shut down as a result of fuel-element failures.

August 1971   The KNK reactor becomes critical for the first time.

October 1971   Last session of the German Atom Commission. It is replaced by the Expert Commission on Nuclear Research and Technology, constituted in December 1971. Contracts are signed for the construction of the 300-MWe high-temperature gas-cooled reactor prototype THTR-300.

December 1971   Brown, Boveri & Cie takes a Babcock & Wilcox (United States) license for pressurized-water reactors. Babcock, Brown Reaktor GmbH is formed as a joint subsidiary.

January 1972   Klaus von Dohnanyi succeeds Leussink as Minister for Education and Science.

August 1972   The KNK reactor delivers the first electricity.

December 1972   The responsibilities of the Ministry for Education and

Science are divided in a Ministry for Research and Technology and a Ministry for Education and Science. Nuclear power is under Horst Ehmke, Minister for Research and Technology.

April 1973   The nuclear reactor departments of AEG and Siemens are merged with Kraftwerk Union.

April 1973   Construction start of the 300-MWe liquid-metal fast breeder prototype plant SNR-300.

February 1974   The KNK reactor operates at full output for the fist time.

May 1974   Hans Matthöfer follows Ehmke as Minister for Research and Technology.

July 1974   The KKN plant is shut down before commerical operation.

September 1974   The KNK reactor is shut down for conversion to a fast reactor.

December 1974   Biblis A starts full-load operation.

June 1975   Brazil and West Germany sign an agreement including orders for two nuclear plants and the transfer of technology for uranium enrichment and reprocessing.

January 1977   Siemens takes over AEG's shares in Kraftwerk Union.

October 1977   The KNK reactor achieves first criticality with a fast core.

February 1978   Volker Hauff succeeds Matthöfer as Minister for Research and Technology.

March 1979   KNK II starts operation at full power.

Appendix B:
Organization Charts of the German Atom Commission

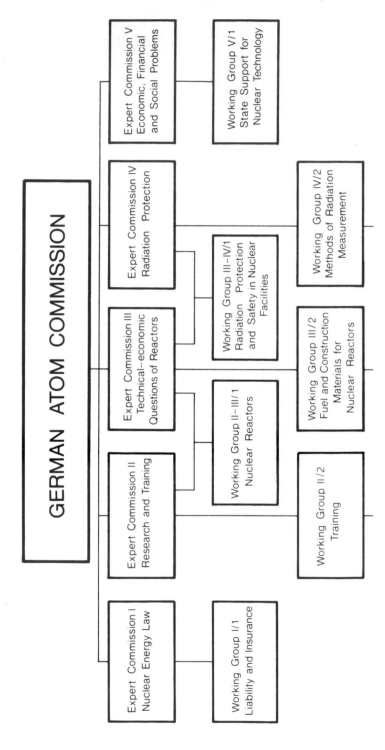

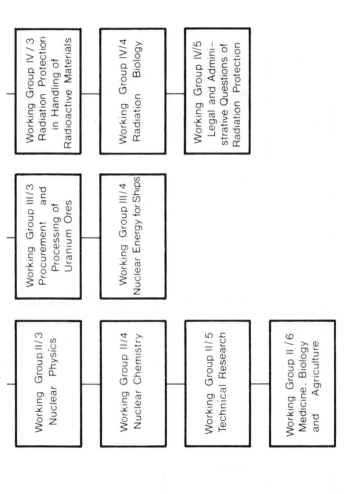

Source: R. Gerwin, *Kernenergie in Deutschland* (Düsseldorf and Vienna: Econ, 1964), p. 171. Reprinted with permission.

Figure B-1. Organization Chart of the German Atom Commission, 1956 to 1966

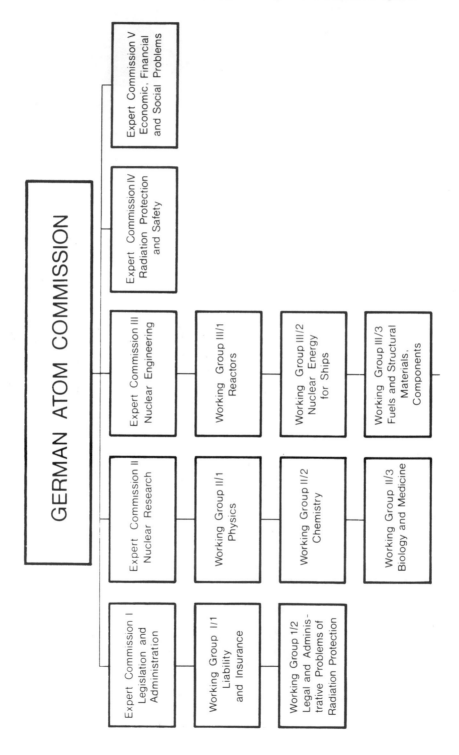

GERMAN ATOM COMMISSION

Expert Commission I
Legislation and
Administration

Working Group I/1
Liability
and Insurance

Working Group I/2
Legal and Adminis-
trative Problems of
Radiation Protection

Expert Commission II
Nuclear Research

Working Group II/1
Physics

Working Group II/2
Chemistry

Working Group II/3
Biology and Medicine

Expert Commission III
Nuclear Engineering

Working Group III/1
Reactors

Working Group III/2
Nuclear Energy
for Ships

Working Group III/3
Fuels and Structural
Materials.
Components

Expert Commission IV
Radiation Protection
and Safety

Expert Commission V
Economic, Financial
and Social Problems

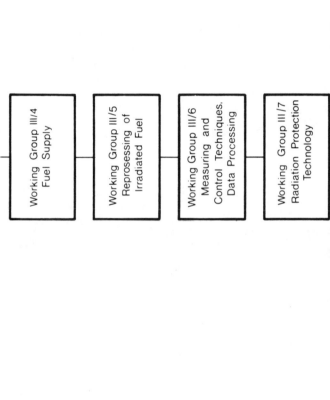

Source: Based on W. Cartellieri et al., eds., *Taschenbuch für Atomfragen 1968* (Bonn: Festland Verlag, 1968), pp. 516–518.

Figure B–2. Organization Chart of the German Atom Commission, 1966 to 1971

**Appendix C:
Survey of Liquid-Metal
Fast Breeder Reactors**

Table C–1
Survey of Liquid-Metal Fast Breeder Reactors

Country (start of FBR program)	Name of Plant	Output		Fuel at Operation Start	Primary Cooling System	Time Schedule (month/year): Start of				
		Thermal (MWt)	Electrical (MWe)			Design	Construction	First Criticality	Full Operation	Shutdown
United States (1946)	Clementine	0.025	—	Pu metal	Loop	1945	9/1946	11/1946	3/1949	6/1953
	EBR-I	1.2	0.2	U metal	Loop	1945	1949	8/1951	12/1951	1963
	LAMPRE	1	—	Pu metal	Loop	1955	1959	1961	1961	1963
	EFFBR	200	66	U metal	Loop	1951	8/1956	8/1963	8/1966	1972
	EBR-II	62.5	20	U metal	Pool	1954	1957	10/1961	4/1965	—
	SEFOR	20	—	Pu/U oxide	Loop	1961	10/1965	5/1969	1/1971	1972
	FFTF	400	—	Pu/U oxide	Loop	1965	1966	2/1980	—	—
	CRBR	975	350	Pu/U oxide	Loop	1969	—	—	—	—
Soviet Union (1948)	BR-2	0.1	—	Pu metal					1956	1957
	BR-5	5	—	Pu oxide		1956	1957	6/1958	7/1959	1972
	BR-10	10	—	Pu oxide						1977
	BOR-60	60	12	U oxide	Loop		5/1965	12/1969		—
	BN-350	1,000	350	U or Pu/U oxide		1963	1964	11/1972	3/1973	—
	BN-600	1,470	600	U or Pu/U oxide	Pool		1968	4/1980	—	—
	BN-1500	4,000	1,500							—
United Kingdom (1951)	DFR	72	15	U metal	Pool		3/1955	11/1959	7/1963	3/1977
	PFR	600	250	Pu/U oxide	Pool		1966	3/1974	—	—
	CDFR		1,300	Pu/U oxide	Pool		—	—	—	—

Country	Reactor									
France (1957)	Rapsodie	24/40	—	Pu/U oxide	Pool	1958	1962	1/1967	3/1967(24 MWt) 1970(40 MWt)	—
	Phénix	563	250	Pu/U oxide	Pool	1964	1968	8/1973	7/1974	—
	Super-Phénix	3,000	1,200	Pu/U oxide	Pool	1971	1977	—	—	—
West Germany (1960)	KNK II	58	18	Pu/U oxide	Loop	1968	1975	10/1977	3/1979	—
	SNR-300	770	280/300	Pu/U oxide	Loop	11/1966	4/1973	—	—	—
	SNR-2	3,420	1,300/1,500	Pu/U oxide	Loop	1972	—	—	—	—
Japan (1967)	JOYO	50/100	—	Pu/U oxide	Loop	1967	3/1970	4/1977	—	—
	MONJU	714	300	Pu/U oxide	Loop	1968	—	—	—	—
Italy (1962)	PEC	120	—	Pu/U oxide	Loop	1969	1974	—	—	—

Sources: W. Häfele et al., "Fast Breeder Reactors," *Annual Review of Nuclear Science* 20 (1970):393–434; R. Rockstroh, "Die Entwicklung schneller Reaktoren," *Kernenergie* 15 (1972,73–77, 213–219; *Design, Construction and Operating Experience of Demonstration LMFBRs*, Proceedings of an International Symposium, Bologna, Italy, 10–14 April 1978 (Vienna: International Atomic Energy Agency, 1978); *Atomwirtschaft*, 1978–1980; *Nuclear Engineering International*, 1978–1980; *Nuclear News*, 1978–1980.

Index

AB Atomenergi (Sweden), 121
Advanced converter reactors, 103–104, 206, 221, 231
Advanced gas-cooled reactor (AGR). *See* Gas-graphite reactor
Advisory committees, 229–239. *See also* Expert Commission for Nuclear Research and Technology; Fast Breeder Project Committee; German Atom Commission
AEG (Allgemeine Elektricitäts-Gesellschaft), 22; boiling-water reactor, 28, 35, 41, 111; commercial orders, 41, 46; commitment of own funds, 47, 101, 121, 127; financial losses, 46; 500 Megawatt Program, 28, 30–31, 35; General Electric license, 22, 25, 44; Kraftwerk Union, 44; light-water reactor, 28, 30; Obrigheim bid, 41; Program for Advanced Reactors, 37–38; research reactors, 25–27; steam-cooled fast breeder reactor, 91, 94, 96, 121, 123–130, 132–133, 226, 233, 238. *See also* Gundremmingen; HDR reactor; Kornbichler, H.; Lingen; VAK reactor
A.E.I.-John Thompson Nuclear Energy Company Ltd. (Great Britain), 26
Agip Nucleare (Italy), 174
AKS association (Arbeitsgemeinschaft Baden-Württemberg zum Studium der Errichtung eines Kernkraftwerks), 23, 30–34, 41
Alkem, 145, 172, 176
Allen, D.H., 107
American Machine and Foundry Atomics, Inc., 23, 26
Arbeitsgemeinschaft Kernkraftwerk Stuttgart. *See* AKS association
Arbeitsgemeinschaft Versuchs-Reaktor GmbH. *See* AVR association
Argentine, nuclear power plant exports, 44
Argonne National Laboratory (United States): Argonaut research reactor, 25–27; fast breeder reactor, 61, 78
Atom Commission. *See* German Atom Commission
Atomic Energy Commission. *See* U.S. Atomic Energy Commission
Atomics International (United States), 103, 104, 107; AKS project, 34; fast breeder reactor, 107; Interatom, 22, 28, 34; organic-cooled reactor, 34, 104; research reactor, 26–27; sodium-graphite reactor, 104; U.S. Atomic Energy Commission, 103, 104

Atom Ministry, 24, 53, 191–193, 225–229, 21–59, 61–88. *See also* Research Ministry; Science Ministry
AVR association (Arbeitsgemeinschaft Deutscher Energieversorgungsunternehmen zur Vorbereitung eines Leistungs-Versuchs-Reaktors e.V., since 1959 Arbeitsgemeinschaft Versuchs-Reaktor GmbH), 23, 30–31, 33
AVR reactor, 30–31, 33, 38, 49; cost, 33

Babcock-Brown Boveri Reaktor GmbH (BBR), 44–45, 47
Babcock & Wilcox (Great Britain), 39
Babcock & Wilcox (United States): Babcock-Brown Boveri Reaktor GmbH, 44; license to Brown, Boveri & Cie, 46; light-water reactor, 107; research reactor, 26–27; spectral-shift-control reactor, 104; steam-cooled fast breeder reactor, 121. *See also* Deutsche Babcock & Wilcox
Badenwerk AG, 42, 63
Baden-Württemberg, 25–27, 62–66
Balke, S., 65
Bayernwerk AG: VAK reactor, 33; Gundremmingen, 39–40, 42–43
BBC/Krupp consortium, 22, 98; AVR reactor, 23, 33, 35–36; 500 Megawatt Program, 28, 30, 35–36; Program for Advanced Reactors, 37–39. *See also* Brown, Boveri/Krupp Reaktorbau GmbH
Bechert, K., 246
Belgium: BR2 reactor, 79, 236; Centre d'Etude de l'Energie Nucléaire/Studiecentrum voor Kernenergie (CEN/SCK), 74; Commissariat à l'Energie Nucléaire (CEA), 144; fast breeder reactor, 69, 74–75, 97–98, 142–146, 227, 243–244
Belgonucléaire, 74, 143, 226–227; collaboration with Alkem, 172; collaboration with Interatom, 142–145; Enrico Fermi Fast Breeder Reactor, 74, 97–98, 172; Euratom, 74–75, participation in the West German fast breeder program, 74–75, 142–145, 172, 174, 233; steam-cooled fast breeder reactor, 121
Berliner Kraft und Licht AG (Bewag), 34
Boiling-water reactor (BWR): AEG, 28, 35, 41, 111; electricity-cost estimate, 111;

About the Author

Otto Keck studied theology, philosophy, and economics at the Universities of Tuebingen and Heidelberg. He did postgraduate work in history and social studies of science at the University of Sussex, where he received the doctoral degree. He has worked as a researcher at the Department of Social Studies of Science of the University of Ulm for the past six years, and in the academic year 1980–1981, he is John F. Kennedy Memorial Fellow at Harvard University, where he does research at the Center for Science and International Affairs, and at the Center for European Studies. He is coeditor of *Technischer und sozialer Wandel—eine Herausforderung an die Sozialwissenschaften* and has published articles in *Research Policy, Energy Policy,* and *La Recherche*.